I0061381

Science in Resistance

The publisher and the University of California Press Foundation gratefully acknowledge the generous support of the Lawrence Grauman, Jr. Fund.

The publisher also gratefully acknowledges the generous support of the Director's Circle of the University of California Press Foundation, whose members are:

Stephen and Melva Arditti
Venetta Campbell
Michael Dear
JAG
Marilyn Lee and Harvey Schneider
Leslie Scalapino - O Books Fund
Susan Magee
Professor Donald Mastronarde and Joan Mastronarde
Sandy Otellini
Alejandro Portes
Sharon Simpson
Catherine Stevens
Lynne Withey

Science in Resistance

The Scientist Rebellion for Climate Justice

Fernando Racimo

UNIVERSITY OF CALIFORNIA PRESS

University of California Press
Oakland, California

© 2025 by Fernando Racimo

All rights reserved.

Library of Congress Cataloging-in-Publication Data

Names: Racimo, Fernando, author.
Title: Science in resistance : the scientist rebellion for climate justice / Fernando Racimo.
Description: Oakland, California : University of California Press, [2025] | Includes bibliographical references and index.
Identifiers: LCCN 2025004977 (print) | LCCN 2025004978 (ebook) | ISBN 9780520420335 (cloth) | ISBN 9780520420342 (paperback) | ISBN 9780520420359 (ebook)
Subjects: LCSH: Climate justice. | Scientists—Political activity. | Climatic changes—Social aspects. | Civil disobedience—Environmental aspects.
Classification: LCC GE220 .R335 2025 (print) | LCC GE220 (ebook) | DDC 363.7/052—dc23/eng/20250531
LC record available at https://lccn.loc.gov/2025004977
LC ebook record available at https://lccn.loc.gov/2025004978

Manufactured in the United States of America

GPSR Authorized Representative: Easy Access System Europe, Mustamäe tee 50, 10621 Tallinn, Estonia, gpsr.requests@easproject.com

34 33 32 31 30 29 28 27 26 25
10 9 8 7 6 5 4 3 2 1

To Matias and Sofia,
so that you know what we did.

Remember the soft place in your heart, that place that you protect,
where you keep your kindness, your care, your fire.
That place from which you first understood that the world needed saving.
From "Climate Activist's Call," by Gregory Manni[1]

Contents

A Rainy Day

It's a cold, steady drizzle, the kind typical of Denmark this time of year. April in Copenhagen. And there's always some wind, so the raindrops hit you straight in the face, like tiny pinpricks.

I am standing in Slotsplads—a central square of the Danish capital—wishing I had brought a warmer jacket. The Christiansborg Palace towers above me. Built and rebuilt after several historical fires, it is a national symbol of resilience and authority. It was once the seat of kings. Today, the palace houses all three branches of the Danish government: the prime minister's office, the Parliament, and the Supreme Court.

Staring at the palace windows, right below the central steeple, I remind myself of why I'm here. I am about to break the law.

I am not alone. There are more than 30 people around me. Some are familiar faces; others I've never met before, but I'm glad they are here. At least half of us are scientists: biologists, geologists, climatologists, sociologists. The rest are people from other walks of life, who came to give us a hand. I can see teachers, doctors, students, and workers from various professions.

We all stand expectant under the rain. We are all about to break the law.

The police are here too. Five armored police vans are dispersed around the square, and a dozen officers patrol the area. They know that we will soon attempt to block the road somewhere near a major government building. They know this because they are following us on social media: several weeks ago, we announced that we would be here. Even though we are about to break the law, we don't want to hide what we are doing.

We live in one of the richest, most prosperous countries in Europe. Danish citizens are ranked among the happiest people in the world. For academics, Denmark is an especially attractive place of work, with plenty of research funding and world-class universities. And yet, here we are: over 30 people—many of us scientists—all about to break the law.

So why are we doing this?

Today, April 6, 2022, is the first global "scientist rebellion": over 1,000 scientists in more than 28 countries are rising up in civil disobedience. We are calling on our respective governments to acknowledge that they have failed to keep their citizens within the warming limit of the Paris Agreement and to drastically curb CO_2 emissions. It's not just Copenhagen—other uprisings are happening in major capitals in five continents: London, Quito, Madrid, Washington, Lilongwe. They are all playing host to their own rebellions: scientists are breaking the law by blocking roads, occupying the headquarters of oil companies, and barging into government buildings. By provoking mass arrests, we hope to catch the attention of the media and the broader public about the urgent need for climate action—attention we have been trying to get for decades via more polite approaches.

We have a briefing to talk about our plan, about what is about to unfold. We evaluate potential risks, check safety protocols, and set up a plan B, in case things don't go as intended. Then, we all start marching. We are carrying a couple of banners, a megaphone, and large cardboard printouts of scientific papers on climate change. We intend to block the intersection in front of the Ministry of Climate,

Energy and Utilities, just a block away from the palace. While preventing traffic from moving, we will read aloud fragments of the latest report by the Intergovernmental Panel on Climate Change, or IPCC for short.

The IPCC compiles the work of thousands of scientists and publishes it in a report that is hundreds of pages long. It is updated roughly twice each decade: there have been six reports since 1990. They are intended to provide critical information to governments and the public about the latest research on climate change. A herculean effort in coordination, the reports are testimony to the volumes of data so far collected about how our Earth system works. They are also beacons of our future: they tell us what lies ahead if governments and corporations continue burning more and more fossil fuels—a future of storms, floods, fires, and crop collapses; a future of famine, mass displacements, and war. António Guterres—the UN secretary general—has just given a statement about the specific report we are about to read. He called it a "code red for humanity."[1]

Wearing white lab coats, we march with copies of the report in our hands, closely followed by the police.

Since early childhood, I remember being fascinated by the people in lab coats, starstruck by the questions they strove to answer. What was the universe made of? How did life on Earth emerge? How could evolution bring so many species into existence, and how could they go extinct in the blink of an eye? While in school, I daydreamed about dinosaurs, meteorites, and DNA. And during the evenings, I dived into the popular science books and magazines my parents would get for me. By the time I graduated from high school, I knew I wanted to be a scientist.

A mixture of luck, privilege, and community support helped me climb the academic ladder, and so, in my early 30s, I became an associate professor in evolutionary biology. I thought my life was pretty much set: carry out research, report on my findings, and teach future scientists about the methods and ideas I had learned along the way. I was determined to become the person my younger self would have admired—a seeker of truth, an explorer, an *expert*.

Then, soon after I had landed my dream job, the climate movement became a global phenomenon. Students in the hundreds of thousands started defying their teachers, skipping class, and marching onto the streets. They couldn't afford to daydream; their futures were being stolen right in front of their eyes. I realized then that my passion for seeking scientific "truths" had blinded me to a much bigger truth. It was not that the world was on fire—by then, I knew this already—but a much more painful truth to face: that those in charge were fanning the flames.

As I write these words, the global average temperature is over 1.2 degrees Celsius higher than it was before the start of the twentieth century.[2] It will be higher by the time you read them.[3] Over 60 percent of climate scientists project that, by 2100, global temperatures will rise at least a further 1 to 3 degrees, if we continue on our current trajectory.[4] These might seem like small numbers, but they entail enormous humanitarian costs. Conservative projections suggest about one-third of humanity will be exposed to extreme and potentially deadly heat by the end of the century, under current policies.[5] Even at just 2 degrees of warming, the IPCC projects up to one-third of the planet's population will be subject to chronic water shortages, while millions more will starve, due to dwindling food supplies.[6] Studies predict widespread collapses in social services and critical infrastructure.[7] Books, papers, and reports are being written about what this unraveling of our societies might look like, and how humans might adapt.[8]

But this is not a problem of the future alone. For many, collapse is already happening, as farming communities, villages, and even whole cities become unlivable due to extreme weather events.[9] The consequences of climate change are today a reality for millions of people, nowhere more than in the Global South. Those being hit hardest are also those least responsible for the growing chaos. In the face of this injustice, we are constantly told to look away. We are all on a "green transition," on a "path to net zero," so keep calm and carry on. Things will look brighter after the next election.

At the same time, we—the citizens, the "electorate," those who can only look at the windows of government from the outside—are kept away from crucial knowledge: knowledge about how much power we actually hold; knowledge about what collective action can achieve and about what it has achieved throughout history; knowledge about how people can come together and challenge those above them. Instead, politicians and corporate leaders tell us to focus on our individual behaviors and trust technology: eat green, recycle, turn off the lights when you leave the room. Some new gadget will come along soon enough and solve this crisis. And it will also solve the next crisis. And perhaps the next. Hell, technology will solve everything. Just keep calm. Just carry on.

Year after year, endlessly repeating climate summits paint a picture of hope, a picture where climate is front and center in the global agenda. But for all the pledges and promises, business as usual seems to be the path of choice. The rate at which governments and corporations are emitting CO_2 continues to rise every year. Indeed, under the dominant economic system, it is meant to rise—the ever-expanding fire an effluent of capital expansion, of unbridled growth, sanctioned and protected by the state. This is another painful truth to accept: *the fire is a feature, not a bug.*

This is why I am breaking the law today: because the law has failed us; because I can no longer sit back and watch as our leaders drive us into the abyss; because my children will ask me one day what I did when I could have done something, and I want to have an answer for them.

We stand in the middle of the street by the Ministry and take turns reading the IPCC report on the megaphone. The rest of us remain silent, holding up our banners and our scientific papers. The rainwater starts seeping into our clothes. And the traffic starts piling up.

The police are shouting, ordering us to move. We stand our ground. They repeat their order, and we remain impassive. So they begin to detain us.

A couple of officers drag me off the road and pin me to a wall. As one of them takes my information, I look over his shoulder: two of my colleagues, Nikoline and Nadia, are being handcuffed and taken into one of the vans. Their lab coats remind me that they are scientists—the first ones in Denmark arrested for climate civil resistance since the global school strikes, arrested for peacefully demanding that politicians act to avert the biggest catastrophe the world has faced. They are the first, but they will not be the last.

Countless reports, articles, and editorials have been written about the science of climate breakdown and the political response to it (or lack thereof). Many more will be written in coming years. But this book is not just about science. It is about the scientists themselves, the *experts* who have been talking about this for years, sometimes decades, and not been heard.

Specifically, this book is the story of those scientists who are recognizing that past tactics have failed and are now choosing new ones. It is the story of scientists who are embracing direct action, the story of underground networks of academics rebelling against the rules of their own institutions and against the system those rules serve. It is the story of ecologists, physicists, chemists, psychologists—all united under a banner of resistance, risking jail to effectively connect with the public about the emergency we are in. It is the story of a radical shift happening today in how academia works—a shift in what it means to be a scientist in the twenty-first century.

This book is also a call to act, a call to become more than just passive observers of the destruction of our world. It is a call to rebellion.

Where Are the Scientists?

"For what good is knowing, unless it is coupled with caring? Science can give us knowing, but caring comes from someplace else."

—Robin Wall Kimmerer[1]

The year 2018 was a year of mass climate protests. Spurred by the words of Swedish student Greta Thunberg, who called it "the time to rebel," students began skipping class each Friday and marching on the streets in at least 270 towns and cities. They were calling out the hypocrisy of the adults in power, who had failed to reduce global greenhouse gas emissions despite 30 years of pledges and promises: "We can't save the world by playing by the rules."[2] By March 2019, more than a million people were joining them on the streets in the first global climate strike.[3]

At around the same time, another movement had also sprung up in the United Kingdom. Extinction Rebellion (XR) was launched in October 2018 and made headlines when it blocked five bridges over the river Thames,[4] as well as multiple road junctions across London.[5] The movement quickly gained momentum and by 2019 had enough support to occupy large areas of the city: Oxford Circus, Marble Arch, and Waterloo Bridge were filled with activists defying orders to disperse,[6] causing so many arrests that the police ran out of holding cells.[7]

XR groups soon emerged in other countries, including Germany, Spain, France, New Zealand, and the United States.[8]

These protests involved breaking the law en masse to push the media and politicians to acknowledge the truth about the crisis and immediately enact emergency climate measures.[9] Unlike the Fridays for Future marches, which emerged organically out of Greta's initial act of defiance, XR campaigns were methodically orchestrated from the beginning. They borrowed tactics from the civil rights and women's suffrage movements and were also inspired by the more recent Arab Spring and Occupy uprisings. As explained in their website, XR had a concrete set of demands and a strategy to achieve them through nonviolent civil disobedience.[10]

I was curious to learn more about them. One day after work, I attended an introductory lecture by two spokespeople from XR Denmark. Held in the scroungy basement of a green NGO, it was unlike any previous talk I had attended. The audience was quite eclectic: there were blue-collar and white-collar workers, a few high school students, and a couple of pensioners. They were all invited to introduce themselves and talk about why they had come. How did they feel about the current state of the climate emergency, about what was being done (or not done) by people in power? Did they share these feelings with family and friends? It didn't look like an informational session—more like a sharing circle for climate anxiety.

Despite some initial reservations, I was moved by the discussion. The XR spokespeople knew their stuff—they had a clear grasp of Earth system science, evidence for biodiversity decline, and theories of political transformation. They could articulate these concepts better than many of my own faculty colleagues. But beyond their clarity, I was moved by the way in which they expressed themselves. These people were not afraid to lay their fear and worries out for their audience to see, and they used their emotions to motivate their activism.

I decided to stick around for a chat with the speakers after the session had ended. I was invited for tea, along with some of the other audience members who were enthusiastic. I mentioned I was a scientist and broached my interest in participating in XR actions. I was struck by their response: "Oh wow! That's great! We really need the scientists to be on the streets!"

That got me thinking. Why *weren't* the scientists on the streets?

SUPPORT . . . AT A DISTANCE

At that time, there was strong support for disruptive protest in academic circles. In April 2019, over 12,000 scientists had signed a letter backing the global school strikes.[11] In Denmark that same year, another letter defending XR's tactics of mass civil disobedience was published by 174 academics.[12] Hundreds of academics in Australia followed suit, with another declaration of support for XR just a few months later.[13] And yet, with few exceptions,[14,15] scientists and scholars were not openly engaging in climate protests, especially in ways that involved disobeying the law. Even those scholars who were encouraging others to participate in direct action were reluctant to become involved in it themselves, to mix science with activism.

British conservation biologist Dr. Charlie Gardner was one of the first to point out the contradiction behind this stance:

> These people have taken to the streets because of what we have told them. The reason students are not going to schools on Fridays is because we have terrified them with the information we have shared. So how can we tell them to act and not be there—shoulder to shoulder—with them?[16]

And it's true: the student marches, the mass acts of disobedience—they were all prompted by what scientists had been saying for decades, by all the international reports on climate collapse, and by the countless papers explaining how bad things will get if we continue with business as usual. Dr. Gardner spoke with me about how he felt attending these protests but not seeing any of his colleagues around:

> It was weird not meeting any scientists at the XR and Fridays for Future protests. I was surprised but also a bit embarrassed. These grandmothers are here, these lawyers are here. But where are the people that wrote the papers that made all the people come here?! I felt that this was wrong, I felt that something I could do was to try to stimulate the scientific community to join this movement. And what more natural than an essay in a

journal? I contacted the only scientist who I heard had self-identified as a member of XR—Claire Wordley, who in 2018 had written a call for everybody to join XR, including scientists. During one of XR's rebellions, I sent her a message asking if she wanted to write something together. Her response was "Fuck yeah!"

In 2019, Dr. Gardner and Dr. Wordley jointly published a public letter in a prominent scientific journal. In it, they compelled their colleagues to consider whether conventional forms of scientific advocacy were sensible at this time, given all the evidence to the contrary:

> Beyond our day jobs, many scientists have also tried to influence policy through available political and economic channels—we have voted in elections, written letters and emails, donated to advocacy groups, and marched in the streets. Many of us have also tried to reduce our personal use of the Earth's resources. However, none of these efforts have worked at the necessary scale. It is time for a new approach.[17]

Scientists were not just absent from public protests—they were very much needed. A more active participation by scientists in civil disobedience could help activists show that they were not just behaving on their own whims—out there causing trouble for trouble's sake— but that they were backed in their actions by those best placed to understand the science. As Wordley and Gardner explained in the same letter:

> Our involvement in popular environmental movements can boost their credibility, change the tone of media reporting, and ensure all members of civil disobedience groups are well versed in climate science, ecology, and other relevant disciplines.

Academic voices (and faces) were clearly in demand. In the United Kingdom, Charlie and Claire's letter precipitated the formation of the group Scientists for Extinction Rebellion, founded by scientists who were openly using their scientific identities in Extinction Rebellion protests. As Charlie told me, "October 2019 was the launch of Scientists for XR. I received an invitation from [two of the group's founders] Emily Grossman and Aaron Thierry to join them for two

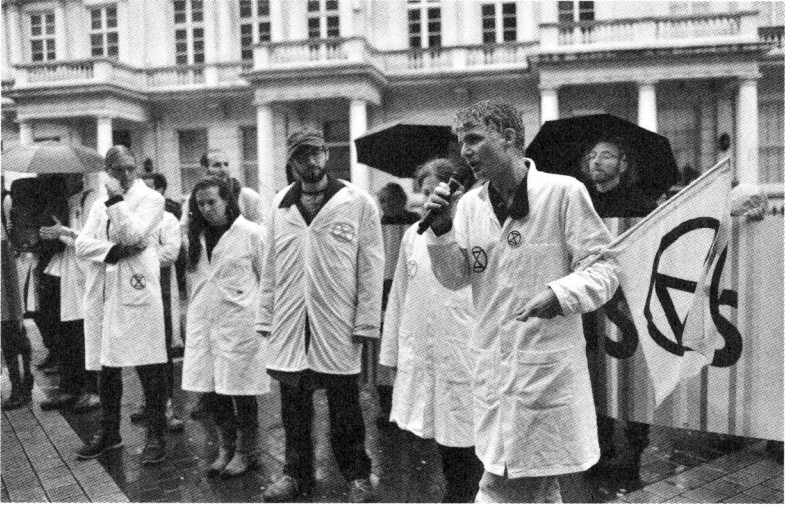

FIGURE 1. Dr. Charlie Gardner speaking at a protest at the Science Museum in London, together with members of the collective Scientists for XR. October 2019. Credit: Louise Jasper.

days of XR events in London. That was the first time that I switched from participating as a citizen to participating as a scientist, and put on a lab coat."

Starting in late 2019 and continuing into 2020, the collective Scientists for XR carried out marches, sit-ins, and protests at major London landmarks—including the Science Museum, due to the funding it was receiving from fossil fuel companies, like Shell, Equinor, and BP. However, Charlie pointed out to me that, at the time, the group was unwilling to feature scientists being arrested:

> In the fall of 2020 there was another big XR rebellion, but it seemed our role in that rebellion was just to be there as a group of scientists in lab coats. I was asking: aren't there any plans to do something where scientists are arrested? And they said, for the time being, no, but there are these two guys who are planning to do an arrestable action, and this is their number: they are doing it this week.

The two guys Charlie was hearing about would eventually become the founders of a new movement—the Scientist Rebellion (or SR for short). And in the course of just a couple of years, the movement they founded would come to transform how scientists participate in the struggle for climate justice. By 2022, Scientist Rebellion was making global headlines: more than 1,000 scientists in over 28 countries were participating in acts of civil disobedience. They were blocking roads, disrupting public events, and occupying corporate headquarters while wearing lab coats, signaling that they were breaking the law *in their role as scientists*.[18] In the United States, a scientist chained herself to the White House fence, demanding the president call a climate emergency. In the Netherlands, Switzerland, and Ecuador, scientists blocked the entrances to government ministries, demanding climate action commensurate with the findings of the latest IPCC report. In Germany, scientists chained themselves with lock-ons while glued to the asphalt on the Kronprinzenbrücke—the bridge linking the Berlin city center to the Bundestag—while calling for a "climate revolution." And in Madrid, dozens of scientists threw red beetroot juice on the steps of the Spanish Congress, alerting the public about the grave danger of climate change and protesting the lack of meaningful action by their government. Among the protesters were internationally recognized researchers, including ecologist Fernando Valladares and social scientist Jorge Riechmann.

Many of these events were met with disproportionate repression. For sitting on the Berlin bridge, geology professor Nikolaus Frotzheim was convicted and fined thousands of euros.[19] And at the time of writing, several of the scientists at the Spanish Congress action are facing criminal charges, with penalties of up to 4 years in jail.[20]

So how did this come to pass?

WHAT THIS BOOK IS (AND WHAT IT IS NOT)

In this book, I will provide an account of the events that triggered this wave of academic rebellion. How did it all start, and expand so quickly across borders? How did scientist-activists start mobilizing in different countries? What was it like for them to engage in direct action for the first time? I will also explore the reasons why scientists are increasingly participating in disruptive forms of activism. What is driving them to leave the comfort of their labs and break the law? What arguments are they using to step out of their conventional academic roles? What kinds of criticism do they receive for doing so? And as the movement evolves, what are some of the tensions that have emerged within it?

Throughout the text, I will draw on findings and ideas from climate science, ecology, political theory, social movement scholarship, social psychology, and philosophy of science. Yet, despite frequent references to the academic literature, this is not an academic textbook or a research monograph. Inspired by the transgressions that scientists themselves are now making, my intention has been to write a text that is (if only a bit) transgressive, something that is academic in nature while also accessible to those outside the ivory tower. Something that—like scientist-activists—crosses the line between the lab and the streets.

I am not a climate scientist—my scientific background is in evolutionary genetics—but neither are most scientists who you'll learn about in this book. We come from many fields, and we welcome all areas of expertise into our movement. We acknowledge that the climate and ecological emergency is a systemic issue: it cuts across climatology, ecology, physics, politics, economics, psychology, and many other spheres of knowledge. My scientific background provided me with a way to understand some of the dimensions of the emergency. It helped me get a grasp of the massive declines in genetic biodiversity the world is going through, at rates never before experienced in human history.[21] However, I found my background quite limiting in many respects, particularly when it came to figuring out what to do with the knowledge I had, how to move forward in the emergency. For that, I turned to other fields. And so it has been with other scientist-activists: as we begin to participate in activism, we come to realize that barriers between scientific silos are not conducive to transformational change. Transgressions need not just be

FIGURE 2. A scientist getting arrested at the steps of the Spanish Congress of Deputies in Madrid on April 6, 2022. Credit: Rodri Minguez.

about breaking the law; they might also entail breaking these barriers and learning to listen to others.

Mixed with the academic scholarship in the book are extracts quoting firsthand accounts by scientist-activists themselves. To get a personal sense of what they think and feel, I conducted interviews with 24 scientists who are currently involved in civil resistance around the world. Some are highly visible in protests, routinely getting arrested and featured in interviews on national media; some are working behind the front lines, helping with action design, planning, and logistics; some are conducting research on civil resistance even as they practice it themselves. Some chose to use pseudonyms, because of repression they are experiencing in their countries. Over the course of these interviews, I asked them why they joined the movement, how their engagement has changed their lives, how they organize for actions in their local groups, and what they think about the future of academia in the midst of the climate emergency. My hope is that this book reads like the voice of not one person, but many.

There are several characters in this book, but really one protagonist: the Scientist Rebellion movement. SR is not the only climate collective fully embracing civil disobedience, but it is the largest one where the "scientific" identity takes center stage. For many scientists, the story of how they came to be involved in activism is also the story of how they found out about SR, how they reached out to its members and received training from them. SR is also the reason why I stepped out of the comfort of my own lab, and it is under the SR banner that I was first arrested. The movement is, of course, still evolving, and it may one day cease to exist. Nevertheless, the ideas developed, lessons learned, and experiences gained by its members will likely shape many struggles in years to come. As we'll see in the following chapters, SR learned tactics from other movements, and other movements are now, in turn, learning tactics from it.

THE STRUCTURE OF THIS BOOK

Throughout the book, I will use actions and events in the history of SR to explore different aspects of the movement and challenges it has faced. I will begin with its inaugural act of disobedience—an action that targeted a high-profile academic society. Featuring interviews with SR's two founders, chapter 1 explains the reasoning behind this action, designed to evince the complicity of scientific institutions in the structural logics of the fossil economy. Chapter 2 then describes how the movement grew in its first year, how scientist-activists in different countries began to connect and collaborate. This chapter is also a summary and critique of the "theories of change" behind conventional scientific behaviors: how most scientists think they can collectively affect the world around them. I contrast these with theories of political transformation that SR espouses, focusing on challenging structures of power rather than only delivering information to them.

Chapters 3 through 7 explore the connections and tensions that exist between scientist-activists and the academic milieus from which they have emerged. In chapter 3, I focus on education—a core component of what it means to be an academic—through the figure of the "teach-in": scientist-activists giving lectures in unconventional locations, displacing the classroom as a form of protest. In chapter 4, I focus on the limited modes of outreach that are today sanctioned by the academy and on how scientist-activists are breaking out of them as well. As they do so, they often encounter criticisms—based on notions of scientific credibility and scientific neutrality—and so I explain and critique each of them in turn. Chapter 5 builds on chapter 4, exploring the deeply ingrained structures of rewards and punishments that limit what are "acceptable" behaviors within the neoliberal university, as well as the ways in which scientist-activists are also challenging them. In chapter 6, I concentrate on the links that exist between academia and the fossil fuel industry and on the campaigns being organized by students and academics to expose and break these links.

Academic activism is also changing the ways in which researchers carry out their everyday work. Nowhere is this truer than in their mobility patterns. In chapter 7, we will meet researchers who are choosing to defy their own institutions—sometimes at the risk of losing their jobs—to avoid reproducing travel behaviors that contribute to climate breakdown. We will also see how activists are confronting the flight industry: blocking airports and private jet tarmacs in order to target the emissions of the ultrarich.

Chapters 8 through 10 explore the dynamics of scientist-activism as it relates to the world beyond the academy. Chapter 8 is an insider account of how Scientist Rebellion members organize and execute an action. What is it like to be involved in the logistical planning? What feelings and emotions go through our heads as we break the law? Chapter 9 is a reflection on what happens when scientists are taken into custody or sentenced to prison. I use this chapter to draw connections between systems of oppression in the modern world (including but not limited to prisons) and the climate injustices scientist-activists

FIGURE 3. Scientist Rebellion at the Ministry of Environment in Quito, demanding the government end its push to increase oil exploitation in the country, April 6, 2022. Credit: Ricardo Morales.

are helping bring to light—systems that act as drivers not only of carbon emissions, but also of mass displacements, war, and genocide.

In facing repression, scientist-activists are relatively privileged. In chapter 10, I reflect on what it means when they join people outside the ivory tower, focusing on environmental defenders and Indigenous groups who are standing against powerful states and corporations to protect their land and livelihoods. Through these struggles, I highlight various pillars of support of extractivism: the cultural penetration of fossil companies into areas rich in hydrocarbons, and the use of "green" discourses to promote mineral extraction.

Finally, the last chapter of the book is a reflection on the future of academia under climate breakdown, in light of the testimonies of scientist-activists in the previous chapters. How is higher education changing as the world passes the 1.5°C threshold of the Paris Agreement? Can we transform our academic institutions to confront the threat of a warming world? I argue that scientist-activism can catalyze a process of democratization in universities, helping academia to rise up to the challenge of the climate and ecological emergency. Academics can learn a lot from activist communities that organize for direct action, and scientist-activists can be a conduit for these lessons to seep into the classroom and the lab.

SOME NOTES ON VOCABULARY

I am an academic, and so—try as I might—I could not help but write a few paragraphs on word definitions and potential sources of confusion that might arise from them. First is the word *scientist*. The *Oxford English Dictionary* defines *scientist*, but online access to this definition requires paying a $70 subscription (valid for 3 months!). And so, I will instead use the one from the free online Wiktionary: a scientist is "one whose activities make use of the scientific method to answer questions regarding the measurable universe. A scientist may be involved in original research, or make use of the results of the research of others." The scientific method is thus at the center of being a scientist, and it is defined (again, by Wiktionary) as "a method of discovering knowledge about the natural world based in making falsifiable predictions (hypotheses), testing them empirically, and developing theories that match known data from repeatable physical experimentation." I would argue that this definition is quite restrictive: sometimes scientists match data from observation rather than experimentation; sometimes scientists are not strictly guided by hypothesis testing, and many knowledge discoveries are not based on falsifiable predictions (at least not directly). Some have even argued that there is no single scientific method, that scientists adapt their methods to the situation at hand.[22] As we'll see in the following chapters, the notion of a "scientific method" and a "natural" (as opposed to unnatural) world that this method is supposed to "discover" are all tied to Eurocentric concepts of what knowledge production is supposed to be. Many would argue that what scientists sometimes claim to discover has been known by people for centuries before the concept of "science" even existed. But we are getting ahead of ourselves. For now, and to avoid confusion, my working definition for a scientist is this one: if you recognize yourself as doing science, and your definition of science doesn't stray *too much* from the one in your nearest dictionary, then, as far as I'm concerned, you're a scientist. How much is "too much" is up for discussion; if in doubt, check with your nearest philosopher of science.

As I said, I am an academic. But not all scientists are academics. And not all academics are scientists. Academics are people who work in academia: universities, research institutes, and other higher-education or scholarly institutions. Many scientists work for corporations, museums, NGOs, or government agencies. I would not call them strictly academics. Conversely, many academics working in the humanities (for example, the study of literature, art, or history) would not call themselves scientists. They might perhaps use the term *scholar*, which could encompass scientists as well.

Why am I making such a fuss about definitions? Well, definitional tensions are also tensions that exist inside social movements. Scientist Rebellion is no exception: Who is a scientist? Do all participants in the movement need to prove they are really scientists? Or is self-recognition as scientist enough? Do we include nonscientist academics or nonscientist scholars in the movement? Not everyone agrees on these questions. Part of the reason for disagreement has to do with the importance of the movement's identity: who are *we*, who can be part of *our* movement? And part of it has to do with risks related to how the movement presents itself to the rest of society: if a member of the movement appears on national TV, criticizing a government official, the public expects the person's definition of what a scientist is to coincide with theirs. If I tell you "I'm a scientist; trust me" but you later find out I'm actually a lawyer or an architect, you might feel duped, and you will certainly not trust me anymore. In general, Scientist Rebellion members rely on each other to be honest about how they identify themselves (inside and outside the movement), and this has worked well so far. All that being said, the words *scientist*, *academic*, and *scholar* often overlap. Throughout the book, I will, for

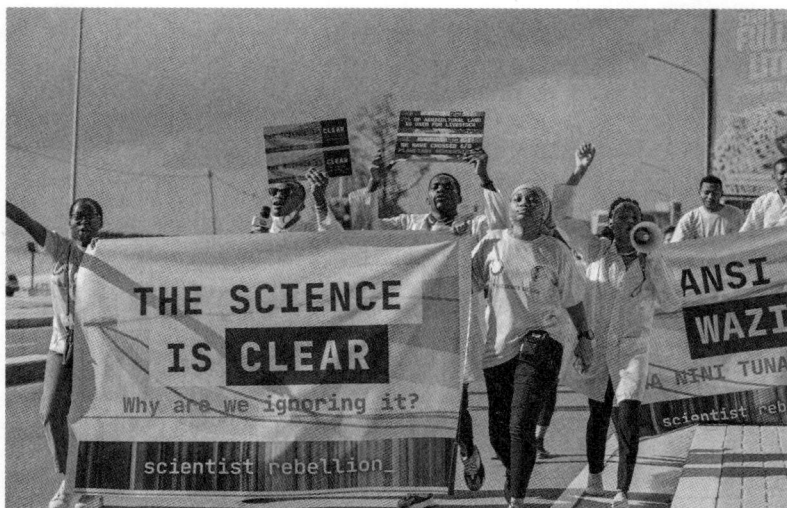

FIGURE 4. Members of Scientist Rebellion protesting on the streets of Dar es Salaam on May 15, 2023. Credit: Gloria / SR Tanzania.

the most part, use these words interchangeably. However, in cases where they clearly do not overlap, I have tried to be careful to use the word that best reflects—to me—the group of people I am referring to.

Another potential source of confusion comes from three pairs of words that appear prominently in this book: *civil disobedience*, *direct action*, and *civil resistance*. Here, I refer the reader to a wonderful study by Oscar Berglund that clarifies and delves into the history of these concepts.[23] In short, civil disobedience comes from the liberal tradition of publicly disobeying the law in the face of injustice, while still being respectful of the rule of law. Those practicing civil disobedience would explain it like this: *We would rather not be breaking the law, but the law (or some aspect of how society is run) is unjust, so we will break the law to highlight this injustice. We are concerned enough about this injustice that we are willing to be arrested, and so we will not resist arrest.*

Direct action, in contrast, comes from the anarchist tradition. Its practitioners would argue along the following lines: *There is an injustice in our society, and we do not trust existing authorities or power structures to address this injustice. Instead, we are going to address this injustice ourselves, working together to achieve this end.* Direct action can include civil disobedience, but it can also include other tactics, like arson or property destruction.

Finally, civil resistance is a concept that encompasses all of civil disobedience and some forms of direct action but puts a strong emphasis on nonviolence. As with the triad *scientist/scholar/academic*, there are overlaps between these three concepts, but also some important distinctions: covert sabotage of an oil pipeline, for example, falls under direct action, but not civil disobedience—people involved in this act might not stick around the site of the crime, waiting for police to arrive and catch them.

Most of the actions Scientist Rebellion performs can fall under any of these three categories. In most cases, therefore, I generally use the words interchangeably. One important aspect of the movement is that actions are carried out openly: members publicly use their identities—as both scientists and citizens—while breaking the law. This means that some forms of direct action (like covert sabotage) are generally not included in the SR repertoire. Like XR, the movement also stipulates that its actions are nonviolent. Whether nonviolence encompasses damage to property and whether such damage is a tactic worth pursuing are both matters of ongoing debate, within SR as well as within the broader climate justice movement. It is not an issue I will dwell upon here, but I refer the interested reader to Andreas Malm's *How to Blow Up a Pipeline* for a recent in-depth approach to the debate.[24]

BETWEEN TWO WORLDS

Documenting the motivations and experiences of scientists engaging in disruptive climate activism is one goal of this book. But it is not the only goal. In the following chapters, I will also try to make a case for the figure of the *scientist-activist* as an emergent political actor in the struggle for climate justice. Scientist-activists are caught between two worlds, while not fully occupying either of them: they are neither just scientists, nor just activists.[25] They have borrowed values and modes of behavior from both the lab and the streets.[26] While activism in academia has a long history preceding Scientist Rebellion, the movement has greatly contributed to cementing the idea that scientists *can* be activists, particularly the kind who engage in actions that involve transgressions (through breaking the law or established conventions).[27] I include myself within this category. I am also a scientist-activist. And so, the story in the following pages is also partly a story about me: how I became involved in the SR movement, while meeting other scientists who were on similar paths, and how these encounters have changed my life.

Beyond documentation and (self-)identification, I wrote this book as a call to act—a manifesto of sorts. As you read it, I hope to convince you, the reader, to also become an activist. Admittedly, because of my own background, some experiences and concepts will likely resonate more with you if you are a scientist. But even if you are not, I intend for this book to also rouse your interest, and, perhaps, your soul. It is arguable whether we need more scientists in this world. But one thing is certain: we need more activists.

CHAPTER I

Acknowledging Failure

"To say one thing and do another—to take one's own word
lightly—cannot inspire trust."

—Paulo Freire[1]

On September 13, 2020, Dr. Tim Hewlett and Mike Lynch-White
meet at 6 Carlton House Terrace, a white stucco mansion just meters
away from Buckingham Palace. The building in front of them is headquarters to the oldest scientific academy in the world: the Royal Society of London. Both Tim and Mike are scientists. Tim is an astrophysics
researcher while Mike is a theoretical physicist. They have agreed to
meet here to deliver a letter to the Society. It reads as follows:

REBEL FOR TRUTH

- We have worked within existing structures of power for decades:
  published papers, advised on policy, spoken to the press; meanwhile, ecosystems have collapsed, emissions have accelerated, the
  future has been stolen.
- There is great value in the work of scientists, but we must
  acknowledge this failure and now engage in new ways.
- Climate scientists have made consistent predictions for four
  decades. In that time our political system—short-termist,
  beholden to funders, media moguls and interest groups, corrupt—
  has proven itself incapable of tackling the emergency.

- Without broad and rapid change, billions of people are likely to die due to climate breakdown this century.
- Those with power must be compelled to act, or they must be removed from power.
- Some changes must be institutional: transforming the many ways in which science relates to industry, finance, war and technology. Other changes require us all to act.
- There is a moral imperative to non-violently rise up against the industries and governments who bring such destruction, in defense of truth and life itself.
- We must lead by example, because when our actions are not consistent with the facts the entire scientific endeavor is undermined.
- If scientists do not act as if their lives depend on political change, why should anybody else?
- Oppose the criminal action and inaction taken by governments and other institutions, the ecocide, the injustice, the construction of genocide.
- In the tradition of Suffragette, Civil Rights, Black Lives Matter and other movements, we are calling on scientists to embody the truth and engage in non-violent direct action.

The presentation of the letter goes viral on social media, but it is not just the text that makes an impact. Mike and Tim nail the letter to the doors of the Royal Society building and throw green paint on its front columns. They then handcuff themselves to the balustrade and wait for the police to arrive at the scene. They are doing exactly what they're preaching: they are engaging in direct action.

Both scientists are arrested and held overnight, along with two other people taking photos.[2] When asked about his motivations, Mike answers bluntly: "If I tell you the room is on fire, and then just calmly sit down and drink a cup of tea, you're not going to believe me, are you?"

Mike and Tim are accusing the institution of not living up to its mission of making science matter to everyone. Yes, the Royal Society publishes hundreds of scientific papers each year and organizes countless conferences, many on climate change. But for all this work,

FIGURE 5. Tim Hewlett (left) and Mike Lynch-White (right) sitting on the steps of the Royal Society of London after tacking their letter to the door of the building. Credit: anonymous (along with Mike and Tim, the photographer was arrested and held in custody for 22 hours without charge).

as emissions keep rising and ecosystems collapsing, the Society has remained impassive. It is as if we had an alert system telling us daily the precise rate and magnitude at which the ground below us is vibrating, yet failing to shout "RUN!" at the first sign of an earthquake. The Royal Society and other scientific institutions might be generating information, sure, but they have not been acting in a way that is consonant with it. In the words of Tim and Mike, they have not *embodied the truth*. Like critical pedagogist Paulo Freire once remarked, they are saying one thing and doing another.

Recalling this action years later, Tim explained to me that this dissonance between words and action was what motivated the choice of venue:

> The Royal Society is allegedly the most prestigious institution of science in the world but is not doing anything at all with the climate crisis, and that's a good example of what's wrong with science today. So it seemed like a natural choice: a way to say "science needs to be better and this is why." Along with the letter, we pasted papers that were actually published by Royal Society journals. We wanted to symbolize that these were their own words. That way, if they wanted to take us to court, we could just submit the papers they had published as evidence of their hypocrisy.

Mike and Tim's message was not directed only at the Royal Society. It was a wake-up call to all academics: a call to recognize that scientific dissemination today is failing—and failing badly. Pick up almost any book, article, or report on climate breakdown these days, and you will find a simple fact inside: the concentration of greenhouse gases in our atmosphere is still rising. Not only that, it is rising at ever faster rates.[3] With increasing emissions, our planet destabilizes more each year, threatening the livelihoods of millions around the world.[4]

It is a daunting fact: for all the conferences, protocols, agreements, and treaties over the past 40 years, modern societies are still actively working toward their own collapse. In 2021, the Intergovernmental Panel on Climate Change (IPCC) reported that global warming is, with high probability, projected to exceed 2°C during this century.[5] More recent estimates point to even higher levels of warming.[6] With each fraction of a degree, the risk of catastrophes also increases. Studies predict a dramatic rise in floods in urban areas,[7] an expansion of infectious diseases,[8] shortages of water due to wildfires,[9] and mass deaths due to extreme heat bouts.[10] We are falling off a precipice, at higher speeds than ever before.

And climate breakdown is only one side of the problem. We are also destroying our living world, on land and in the sea. The past 50 years have witnessed the extermination of living beings at a scale only comparable to mass extinctions. Up to a million species are now at risk of disappearance, including approximately 10 percent of insect species[11] and 25 percent of plant species.[12] And we depend on many of these for our food: insect pollinators and keystone plants keep ecosystems in check; they create the conditions that enable farming and grazing. Plus, the atmosphere we are disrupting is the same atmosphere that has allowed us to actually *grow* that food for thousands of years. Couple these facts together, and the overall picture looks bleak: the IPCC warns of "severe impacts on crop production" in coming decades.[13] By 2050, somewhere between a tenth and a third of global crop and livestock area is expected to become climatically unsuitable.[14]

But this is not just a problem for future generations. It is everywhere around us today. In 2022, deadly flash floods hit Pakistan, caused by monsoon rainfall 5 times heavier than average. They inflicted more than $10 billion in damage, affecting over 33 million people and causing over 1,000 direct deaths. Over 1 million houses were damaged or destroyed, and 2 million acres of crops were ruined. This unleashed a famine in a country where a large portion of the population already had little access to food before the disaster.[15] Meanwhile, in Europe that same year, extreme heat waves led to wildfires in Spain and Portugal, causing droughts and crop failures unlike anything seen in the past 1,000 years.[16] Dried-up rivers in France and Germany caused power shortages across the continent, affecting millions.[17]

Scientists knew this would happen. Indeed, they have been issuing warnings about the dangers of climate breakdown for decades. Why, then, has our political system, as Tim and Mike were claiming, proven itself "incapable of tackling the emergency"? Perhaps it is not more information that is needed, but something else.

THE GEARS OF POWER

We are causing the climate to change. *We* are driving species to extinction. We hear this all the time—in the news, in corporate advertisements, even from friends and colleagues who often remind us of our environmental footprint and of ways to lower it. Yet the amount and nature of the blame for the state of our world is far from being evenly distributed. To understand our predicament, we need to delve deeper into what we mean by *we*.

Let's focus on CO_2 emissions. When one hears "we are emitting CO_2," one could imagine we are all doing it at roughly equal rates. Yet the richest people in the world, the top 10 percent in terms of wealth, are responsible for roughly half the world's emissions when you account for the consumption (of goods and energy) that propels that CO_2 to be generated in the first place. Within this elite, there is an even higher-emitting superelite—the top 1 percent, responsible for roughly 15 percent of the world's CO_2 by consumption. This is equivalent to more than half of the emissions of the bottom 50 percent.[18]

Clearly, not everyone is equally responsible for climate breakdown. And responsibility is not just about unequal patterns of emissions, but about the economic relations that keep societies dependent on *ever-increasing amounts* of emissions. While most of us consume fossil fuels—our rising electricity and fuel bills a constant reminder—there are only a few who stand to gain from the expansion of this consumption, only a few who increase their capital for every additional ton of CO_2 that gets pumped into the air. Between 1985 and 2018, as global temperatures rose by almost 1°C, fossil fuel producers made 30 trillion dollars in profits.[19] Most of us have never had a say in this economic arrangement. It has been foisted on humanity by a tiny elite, what Marxists call the dominant class. As Andreas Malm explains in his book *Fossil Capital*, "if humanity as a whole drives the locomotive, there is no one to depose."[20] He then continues:

> The succession of fossil-fueled technologies following steam—electricity, the internal combustion engine, the petroleum complex: cars, tankers, refineries, petrochemicals, aviation . . .—have all been introduced through investment decisions, sometimes with crucial input from certain governments but rarely through democratic deliberation. The privilege of instigating new cycles of burning appears prima facie to have stayed with the class in charge of commodity production.[21]

If the way our economy runs is not a matter of democratic delibera-tion, then information can only take us so far. You might get scien-tists to inform 99 percent of people about the danger of more oil rigs, more pipelines, more plastics, and more agrochemicals. But, as long as the 1 percent remain in control and stand to gain from them, the matter remains moot. So it is power—not information—that we should put our focus on.

In a recent study, Isak Stoddard and colleagues spotlight the main players of the fossil economy—those who have a vested interest in the continued expansion of oil, coal, and gas.[22] This power "core" includes not only companies that extract fossil fuels (like Shell, TotalEnergies, Exxon, and Aramco), but also the industries that rely on them for com-modity production and transport (think GM, Amazon, H&M, Cargill). It also includes the banks that finance them and the government elites that protect them. The core relies on mechanisms that bolster its power and legitimize it in the eyes of the public. Among these are the ideolo-gies that validate its internal logic, for example, the idea that infinite capital expansion and economic growth are possible on a finite planet.[23] More perniciously, they include the idea that expansion and growth are necessary: the only way to achieve a prosperous society for all.[24]

Other ideas in support of the status quo have to do with membership in what Malm calls "the class in charge of commodity production"—the people who get to decide what and how much gets produced in our society and how energy is harnessed. This is why, even as renewable energy infrastructure—like wind farms and solar panels—has become cheaper than new oil, coal, and gas plants, financial capital and state subsidies keep being pumped into new fossil energy ventures.[25] The dominant class profits because fossil fuels are restricted in space. Who-ever owns a coal mine or an oil rig gets to reap the rewards from its bountiful outputs. In contrast, the sun shines on everyone's roof. Renewable energy is much harder to appropriate and commodify, harder to sell (for a fee).

Isak Stoddard and colleagues additionally highlight the social imaginaries, or mental pictures, in support of fossil capital: what we envision our societies could become. It is in the core's interest to police and shape these imaginaries. Visions of the future remain encased in technocratic scenarios filled with fossil-fueled space exploration, energy-intensive artificial intelligence, and petrochemical food "solutions," while the basic tenets of our economic systems remain unchallenged.[26] In all these futures, capital still controls how we live, how we eat, and what we get to do: Mars colonies, brought to you by SpaceX; flying cars, powered by Shell; a world of technological possibilities, just make sure to pay at the counter.

The constriction of our imaginaries also extends to the lives we aspire to have. Today, the advertising industry elevates high-carbon, high-consumption activities to marks of status and means to freedom in the eyes of the public—think SUVs, fast fashion, and high-tech gadgets.[27] This mode of living is made to represent a universal aspiration, a path to the "good life."[28] Its impact on people and planet is downplayed or made invisible. iPhone ads never feature the sweatshops that produce them, or the cobalt slave mines from which their circuitry is sourced, or even the CO_2 emitted before they finally land in your nearest Apple Store. Out of sight is out of mind.

AND THE BAND PLAYED ON

But where do scientists fit into this picture? Of course, they are not in the power core. They are not steering the ship of the global economy. Instead, they provide information about how the ship is doing and where it is going. They give us ideas and narratives that help us understand how the world works—in a sense, their own form of ideology, grounded in collected data and scientific reasoning. Scientists also shape our imaginaries. As the Royal Society letter remarked, "climate scientists have made consistent predictions for four decades," presenting possible scenarios to the public about what is to come if we continue burning fossil fuels.

Yet, Tim and Mike were pointing out that scientists are not acting as if the ship needs to change course; in a way, they are also bolstering the logic of the power core, even while they alert us of its destructive effects. In our interview, Tim Hewlett explained:

Academia is central to all sorts of industry, and to the legitimacy of government and its choices. Scientists and scientific institutions are important for maintaining the establishment—for maintaining our particular economic and social order.

Scientists are like a violinist playing a beautiful concerto as she watches the back of the auditorium catch on fire. Instead of interrupting the music and loudly shouting that people need to get out, she continues playing, unwilling to cause a stir. Mike and Tim's action was a way of changing the tune scientists are used to playing. They were warning about the fire in a way that got people to listen.

So what are the tunes that scientists usually play? One way that scientists conceptualize and communicate their work is through so-called models. These are statistical tools that allow us to project present conditions about the world into the future—given what we know today, what is likely to happen 10, 20, or 100 years from now? Climate models, for example, provide us with predictions about the behavior of our atmosphere. Ecological models can tell us how ecosystems might respond to increased temperatures or changes in land use. Models also inform the decisions we make in the present, and they inform how we advise policymakers.

So far so good. The problem lies in the assumptions these models make, what they take for granted. Some of these assumptions are stated, others are not. The stated assumptions are the easiest to untangle. Take the models used by international bodies, like the IPCC, to predict how much the world is set to warm. Over the past 15 years, the models the IPCC has used to project into the future (integrated assessment models, or IAMs) have become extremely reliant on the future existence of so-called negative emission technologies, or NETs. These technologies involve absorbing already-emitted CO_2 via artificial processes that remove CO_2 from the source (such as a cement factory) or even directly from the air. NETs then turn the CO_2 gas into a more compressed form, which can be used for other purposes or stored in the ground.[29] Most IAMs today assume large amounts of these technologies will exist at huge scales by the end of the century.[30]

Yet, at the same time, NETs are unproven at the scale needed to pull us away from the most dangerous climate scenarios. On top of that, they involve massive costs in terms of both ecosystem degradation[31] and energy input.[32] To put this into perspective, there are currently 26 commercial CO_2 capture facilities operating worldwide. Combined, they capture approximately 40 million metric tons of CO_2 each year.[33] In contrast, the world already emits over 36 billion metric tons of CO_2 per year: roughly a thousand times more than what is being captured.[34]

NETs are not only inefficient; they are also dangerous red herrings. They have been promoted by governments and fossil fuel companies because the (unproven) potential of NETs allows them to deflect attention away from the real issue: the emissions themselves.[35] Instead, the public is told to put their faith in the promise that NETs will eventually work at the scale needed. John Kerry, the US climate envoy in the Biden administration, conveyed this sentiment well in his 2021 interview with the BBC: "I'm told by scientists that 50% of the reductions we have to make by 2050 or 2045 are going to come from technologies we don't yet have."[36] In other words, business as usual can proceed apace because scientific prophecies will eventually come true. Thus, NETs enable polluters to indefinitely postpone a problem that urgently needs to be tackled today through massive decarbonization.[37] This makes the problem worse every year, encouraging scientists to assume ever-larger quantities of CO_2 capture in their models. We keep running toward a moving target, which keeps getting farther and farther away.[38]

Those are the assumptions that are stated. The ones that are not stated are even more problematic. Models that natural scientists use often pay little attention to the social dynamics of human societies: they try to steer clear of politics. Indeed, natural scientists can go to great lengths to create the impression of being "apolitical," stating their conclusions so that they appear—on the surface—to be valid regardless of how society is organized. Yet, by doing so, these models support the state of power relations as they exist today. For example, IPCC future scenarios (implicitly) assume that the standard of living in Global North countries will remain higher in the future relative to countries of the Global South: countries in the North are not expected to fall in relative life quality, when compared with the South. Thus, rather than aspiring to a future where every life is valued equally, regardless of citizenship, the IPCC perpetuates present-day injustices into the future.[39]

Stated more simply, it is as if I were emitting 10 times as much carbon as you are and then told you that we all must make sacrifices for the climate—we will all reduce our overall emissions, sure, but moving forward, I will still get to emit 10 times as much as you. This is a political decision: one that holds me proportionally less accountable for emission reductions than you so that I can retain my relatively higher standard of living (which I got by polluting more than you ever did). Under the appearance of statistical rigor, a political decision is made, acquires a veil of scientific validity, and is passed off as apolitical.[40]

NO THINGS HELD CONSTANT

There are also strong assumptions made in the models we use to conceptualize our social and economic relations to the rest of the living world. I recently attended a talk where a conservation biologist claimed that "conservation works," citing as evidence that there would have been more species extinctions in the past without conservation efforts. He was referring to a paper arguing that, all other factors held constant, the extinction rates of birds and mammals since the early 1990s would have been 3 to 4 times greater without conservation programs.[41] After the talk, something kept buzzing in my head—something that was missing in this comparison. Sure, *all other factors held constant*, conservation efforts have helped. Yet, why do we *have* to hold other things constant? What about human behavior, and human relations? Ongoing mass extinctions are not a natural product of circumstance; they are constructed by the way we organize ourselves—into a capitalist system that is driven by the exploitation of nature and people so that a tiny group of people can increase their capital. Dr. Charlie Gardner, whom we met in this book's Introduction, drew an analogy for me, to highlight the madness behind this approach:

> I can't pinpoint the time when I first realized this, but it was obvious early in my career that conservation efforts are never going to be enough. They don't address the drivers, the causes of biodiversity loss. When I was young, I remember this picture of a bulldozer running over a landscape. Conservation is like running behind the bulldozer, trying to put things back together, but someone has to be in front of the bulldozer, putting a stop to it! Yet, in conservation biology, our job is to be behind the bulldozer, not in front.

So while conservation efforts have some effect, they will always be insufficient as long as human societies continue organizing themselves around elite profit making and mass resource extraction. In one of his papers, Dr. Gardner and fellow ecologist Dr. James Bullock illustrate this point further, showing that the amount of money spent on global protected areas is approximately equal to the total spent on beard-grooming products around the world.[42] In other words, we spend more on buying razors than governments do on preserving the living world. And if nobody stands in the way of the bulldozer, it will just keep advancing.

Like biodiversity loss, climate breakdown has also been predominantly tackled through the lens of capitalism—from the back, not the front, of the bulldozer.[43] In policy discussions about climate mitigation, the basic principles and assumptions of capitalist economics are rarely challenged, even in the face of overwhelming evidence about the *need* to challenge them. This has led to outrageous statements by economists, unwilling to admit that their models have little connection to reality. For example, Nobel prize–winning economist William Nordhaus studied the present-day relation between GDP (a measure of the market value of all goods and services produced in a given region) and mean temperature across different US states.[44] He was aiming to use this simple relation to predict what a warming world would look like, under different temperatures. From his analysis, Nordhaus famously concluded that the "optimal" scenario for the Earth would be a global temperature increase of 4°C by 2150.[45] Growth could keep going unabated for many more decades to come, helping humanity to adapt to the ever-warming world.

Most climate scientists agree that this is lunacy. It's not just that under a 4°C scenario, with extreme weather events increasing in frequency and magnitude by several fold, "the limits for human adaptation are likely to be exceeded in many parts of the world," as climatologist Rachel Warren puts it.[46] It's also that at a level of warming that severe, multiple planetary "tipping points"[47] would be transgressed.[48] These tipping points are critical thresholds of our climate system: passing them can cause irreversible changes in the way the Earth works, precipitating further warming beyond human control. For example, the Amazon rainforest is currently absorbing CO_2, thanks to an abundance of lush carbon-hungry trees in the region, but it is dangerously close to transitioning into a dry savanna environment, as a consequence of global warming. As the forest disappears and the quantity and magnitude of fires in the region increase, the Amazon may become a net source of CO_2. If transgressed, this and other tipping points could plunge us into a deadly "hothouse" Earth scenario, unlike anything humanity has ever seen.[49] Food supplies worldwide would dwindle. Yet in Nordhaus's analysis, a 4°C increase would lead to a "trivial level of damage."[50] One can only guess how the collapse of global food production could count as "trivial" under this form of economic thinking.

Yet, unlike Nordhaus, we don't have to treat the assumptions of mainstream capitalist economics as sacred and unchangeable. Human societies can be organized in many other ways, so as not to prioritize growth and short-term profit above the prospects of a livable planet. Indeed, humans have organized themselves in many other ways for millennia—before the emergence of capitalism—in societies centered around different types of collective social governance institutions.[51] Many of these societies were complex and could support high population densities while still respecting the natural environment that nourished them.[52]

Many alternatives to the capitalist economic framework already exist. Among them, the "degrowth" school of thought posits that we can achieve socially desirable outcomes—like universal access to food, water, health, and education—within ecological limits, as long as we abandon the idea that infinite economic growth is possible, or even desirable, on a finite planet.[53] Other related approaches and philosophies centered around sustainable ways of life—"sumak kawsay / buen vivir,"[54] "ecological swaraj,"[55] "social ecology,"[56] "doughnut economics,"[57] to name a few—are also becoming popular today, or they are being reinvigorated after decades (or even centuries) of being pushed to the sidelines. It is becoming more and more obvious that capitalist economics is not only not a solution, but the source of our problems.[58]

So to summarize, the models that researchers use to predict the future delimit a particular set of decisions that can be made moving forward. And model assumptions ignore alternative ideas of how we can organize ourselves as a society. These ideas can lead to a downscaling of our consumption and production patterns, and they could save us from the most catastrophic climate scenarios.

The problem with models also has to do with their fixation on the future, at the expense of the past. As decolonial scholar Amitav Ghosh explains, scientific models tend to blind people from the reality that climate change is not just a story about the twenty-first or the twenty-second century, but a story about what came before, about the past injustices that made it all possible:

> The methods of scientists, in particular, have profoundly shaped the way climate change is imagined and thought about. Take, for instance, a procedure that is in widespread use among climate scientists: that of using models to make projections in relation to a future date. Using this method a scientist might predict, for example, that sea levels will rise by so many meters over the next thirty, or eighty, or one hundred years. An unintended consequence of this practice is that it reinforces a perception that abstracts climate change from the past and projects it in the opposite direction, toward the future. . . . the method also casts a long shadow outside the sciences, making it easy to lose sight of the fact that while an event may be unprecedented, the human activities that have left their fingerprint on it may actually be embedded in long and enduring patterns of history.[59]

The historically carbon-intensive activities of the wealthy few—let alone the origins of a dominant class—are not questioned in these models. Nor are the patterns of emissions that led to this point in time, driven by the appropriation of nature and labor by the Global North from the Global South—a process that enriched the former at the expense of the latter. And thus, the models not only perpetuate the rise in emissions, but also fail to address the historical injustices that allowed them to exist in the first place.

A FAILURE TO ACT

Model assumptions are just one way in which scientists have failed to question dominant economic systems. Mike and Tim showed that there is another way in which scientists have failed—not just through their statements, but through their actions. In 2021, a global poll surveyed more than 11,000 people across 28 countries, revealing that 68 percent of the public trust what scientists say about the environment "a great deal or a lot"—among the highest of any profession.[60] Yet, by sticking with conventional behaviors in the face of catastrophe, scientists have not fully honored this trust. They have not acted (in the words of Greta Thunberg) "as if our house is on fire."[61] Tim Hewlett explained:

> If scientists don't act as if our lives depend on resistance, then why should we expect others to do so? The truth is that fossil fuel capitalism can only exist with the tacit or explicit support of scientists or of academia. Academia is a fracture point, a place of weakness you can target to help bring the system to end sooner rather than later.

Admittedly, scientists do not like to advertise their own failure. Many like to think that they are the world's first responders—the people best positioned to alert the public in times of crisis. They are trained to provide solutions, *breakthroughs*, to believe that they and they alone can help people understand their problems better and improve their lives. They are also trained to sell themselves, and to do so in the best possible light. The European Research Council—Europe's most prestigious research funding agency—repeatedly states "excellence" as its sole criterion for giving money to researchers to fund their projects.[62] Universities worldwide encourage their employees to be "competitive," to do and say things that elevate them above others.[63] Thus, scientists are constantly told to promote their own success over that of their peers. It is hardly surprising then that it is so difficult for them to admit failure—particularly a failure of planetary proportions. They want the audience to keep listening to their concerto, the one they've been training to play their whole lives, even as the fire continues to spread across the hall.

Of course, scientists can still honor the trust that has been placed in them. Indeed, this is what those engaging in civil disobedience are trying to do.[64] As we'll see in the next chapter, many of them are finally starting to change their tune. They are recognizing their failure and adjusting the way they engage with the public.

CHAPTER 2

Scientists of the World, Unite!

"If the injustice . . . is of such a nature that it requires you to
be the agent of injustice to another, then I say, break the law.
Let your life be a counter-friction to stop the machine."
—Henry David Thoreau[1]

Soon after their visit to the Royal Society, Mike and Tim started
working on getting other scientists to join them, to participate in civil
disobedience. They made full use of the tools at hand. They exploited
the fact that scientists are well connected internationally and usually
have publicly listed contact information. Tim explained during our
interview:

> Early on, we used social media and email. We scraped the email lists of
> well over 100,000 academics around the world so we could spam them
> and tell them, "This is what we've done. Do you want to be involved?" A
> lot of people were angry at us for spamming them, of course, but some
> were curious. We started having multiple introduction sessions weekly,
> talking about scientist activism. Anyone who had questions or wanted to
> know more, we'd be there to talk to them. We also had plenty of time to
> learn about where each of us were and how we could support each other
> from afar. Once you have a small dedicated team of 2 or 3 people in a
> local area, you can do anything!

Thus, several scientists began to follow in Mike and Tim's footsteps.[2] In
late March 2021, two German researchers—biochemist Nana-Maria

FIGURE 6. Dr. Nana-Maria Grüning pasting scientific papers about the ongoing sixth mass extinction at the Federal Ministry of Food and Agriculture in Berlin. Credit: Johannes Hartl.

Grüning and environmental scientist Kyle Topfer—pasted papers about climate breakdown on the walls of the Federal Ministry of Food and Agriculture in Berlin, to highlight the emissions caused by industrial animal agriculture and monoculture cultivation. In Spain and Switzerland, scientists covered the walls of universities and research centers with papers as well, symbolizing that academic institutions should be in emergency mode. In Mexico, Australia, the United States, and several countries in Europe, scientists went on hunger strikes for days. A software engineer in Djibouti even spent over 2 weeks without eating food, to protest against the effects of fossil emissions in his country.[3] This was the first international series of actions by Scientist Rebellion: more than

100 scientists in about a dozen countries took part in civil disobedience. In the United Kingdom, Mike Lynch-White even upped the ante, locking himself to the gates of 10 Downing Street—the official residence of the British prime minister. It took a specialist police team and a dozen officers to cut him away, after 3 hours.[4]

A series of 4 days of global disruption culminated with Mike and Charlie Gardner carrying out an action at the News Building in London: the heart of Rupert Murdoch's media empire. They pasted scientific papers on the building and threw paint at the front entrance, calling on journalists to report on the crisis with the urgency it deserves, and to stop misleading the public. Both scientists were arrested.[5] Charlie recalled:

> Before the action, I had this call with Tim and Mike. Mike didn't hold back as to what my responsibilities were. I came to see that because I had written that paper in 2019 and had unwittingly established myself as a spokesperson for academic activism, I had a responsibility to also get arrested. Otherwise it would be an abdication of leadership. Mike made me feel that responsibility. We put together a plan to do something at News Corp. I was really really nervous. To be honest, I was stepping well outside my comfort zone. I pressured myself into it. I chose the papers I would paste. I even had a practice at home, pasting one onto a window. I ended up leaving that paper glued to my window for years afterwards! Then I met with Mike in London. He had this big tub of pink paint, and he told me, "We won't get arrested if we just paste papers, so I brought some paint just in case." I pasted the papers, and Mike threw paint on the walls of the building.

In this first wave of scientist rebellion, those acting made sure people took photos and video of what they were doing. They sought to inspire their colleagues to come off the sidelines and add to the momentum. Looking back, Mike recognizes that this helped scientists who had never been involved in activism to change their minds: "Witnessing others like them, they realized they too could do *that*." Those curious might push themselves to come to an information meeting, where they were welcomed by a community of people that talked and behaved like them. They were also provided with resources to act wherever they happened to be.

Yet, Mike stresses, there was another important factor in the movement's early expansion: the fact that actions happened in "waves" of rebellion in which multiple country chapters acted simultaneously. The wave in March 2021 would be followed by another one in April 2022 and yet another in October that year. Aided by secure forms of online communication, SR actions in different countries were coordinated so as to happen on the same day, or at least in the same week:

> I am confident this explains why SR grew internationally so quickly. Extinction Rebellion grew abroad because it was massive in the UK and people wanted to replicate that success. In contrast, SR was tiny when it was already on three continents. But that fact, of acting across borders, was incredibly motivating. People acting in Madrid knew that on the same day, there were people acting on many other continents. I also noticed that every group would look on with admiration at others and be motivated by their successes. So whilst SR Germany, for instance, might be struggling with group dynamics or something, they then see that SR Ecuador just pulled off something incredible over there. And, crucially, while staying a little naive to the problems they of course also faced. There was then a very healthy "competition" between groups: "Yes it's super hard to get scientists to act, but hell, if that country can do it then maybe we can." That belief was very powerful.

Thus, because of simultaneity, the "others like me" effect that inspired individuals to join the movement also pushed country chapters to organize effectively for action.

Of course, there was a lot of work to be done in convincing scientists to break the law. Elena—one of the SR members participating in the uprisings in Spain—remembers the first planning meetings:

> There was initially some confusion about the role of scientists in these actions. Until then, some academics had been collaborating with Extinction Rebellion, but just going to the actions. The common thing was for a scientist to stand on a pedestal talking, while everyone else was getting arrested. I remember saying: "No. No. No. This time, no pedestal!"

Indeed, for SR's founders, arrestability was a central aspect of actions by scientists. This was a point of contention with other scientist-activist groups that existed at the time, which supported civil disobedience but were reticent to feature scientists getting arrested. In Mike's words:

> I think it was absolutely crucial that scientists in actions were arrestable. The group Scientists for XR had been around for a while already in the UK, but barely grown because they weren't putting any skin in the game. They still had the view that scientists are too important to be arrested, and that they would lose credibility. Saying that we face a climate apocalypse but then following it up with saying that a night in a cell is too high a price completely undermines your point. Very early on, we had this debate, and it was only through SR's success that we won that argument.

In the United Kingdom, Scientists for XR ended up adopting tactics similar to those of Scientist Rebellion, organizing events where scientists would be arrested as well. In other countries, like the Netherlands, Scientists for XR rebranded themselves as Scientist Rebellion and began focusing on arrests as a major component of their actions. After this initial wave of expansion, by October 2021, Scientist Rebellion chapters were operational in over 20 countries, all engaging in some form of civil disobedience or direct action. An international network of scientist-activism had been born.

CLIMATE REVOLUTION, OR WE WILL LOSE EVERYTHING

SR's rapid growth inspired its members to meet their counterparts across borders, to see in person those who had only been, until then, faces on a screen and, perhaps, join forces with them. After a series of online meetings, members of SR from Spain, Germany, France, the United Kingdom, and Sweden decided to meet in Glasgow for the 26th Conference of the Parties (COP26) of the UN Framework Convention on Climate Change. As they had been doing in their respective countries, they wanted to show the public that politicians had failed them; in fact, they were calling the entire COP process "a failure."[6]

On November 7, 2021, under a heavy rain, eight scientist-activists stepped onto Glasgow's George V Bridge, which feeds major traffic into and out of the city center. They had lab coats under their clothes. Rapidly pulling them out and putting them on, they formed a human chain, blocking traffic across one of the bridge lanes. They also had an actual chain wrapped around their necks, physically linking them all together. Due to the conference, the city was filled with police, and so a few officers were already patrolling the bridge. As the cars started to honk at the blockaders, the police quickly moved to the site of disruption. While they tried (and failed) to separate the chained scientists, another group of scientist-activists quickly moved into the other car lane, doubling the length of the chain.

The bridge was then fully blocked, entirely by scientists. Momentarily, the police were clueless about how to move them: the chain had bought the scientists some time. As police reinforcements were called in, the blockaders pulled out a long banner extending from one end of the chain to the other. It read: "Climate revolution, or we will lose everything."[7]

The Glasgow action led to the first mass arrest of scientists for climate civil disobedience. In total, 15 scientists were taken into custody. Tim Hewlett explains what it was like to plan the event, train for it, and be arrested with so many colleagues:

> The feeling was brilliant. We had a good group atmosphere. People were determined and willing to work hard. It was certainly stressful, because it was the biggest single action we had organized, and a lot of logistics went into it. There were at least 10,000 extra cops in the city, so it was not easy! We had to train, we had to supply materials, we had to prepare people for dealing with police and the British legal system, all of this in just a few days. It was a lot of stress, but extremely rewarding. And it ended up being a very binding event. I'm still friends with some of those people that I met for a few days a couple of years ago.

Charlie Gardner also recalls the event with great fondness:

> That was my most memorable action. It felt like an exciting time. We had people from all over Europe come together in Glasgow to do this audacious thing that we'd planned. We pulled it off and it was fantastic. On a personal

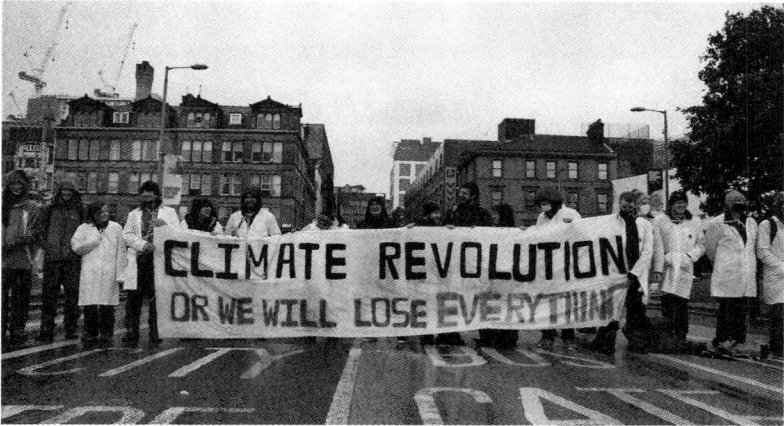

FIGURE 7. The blockade of the King George V Bridge in Glasgow during COP26. Credit: Mar Sala.

level, it was the most fulfilling and rewarding thing that I've ever done as an activist. I've never felt part of something in such a way. It was magic.

The human chain had thus created international bonds of solidarity, which would prove very useful in years to come.

FAULTY THEORIES OF CHANGE

These new acts of rebellion set a new precedent for many scientists, affecting how they envisioned they could act with respect to climate breakdown. For those participating in them, the actions also changed how they talked about the emergency with their own colleagues, friends, and family members. Scientist-activists became bolder in their tabletop conversations—having faced an armed police officer or a honking truck driver face-to-face, they now found it not so stressful to face a furious uncle during dinner. Inevitably, the new scientist-activists were met with counterarguments, many of which had been used to criticize members of other civil disobedience movements in the past:

- *I agree with what you are saying, but your methods are flawed.*
- *You're just making people angrier, and driving them away from your cause.*
- *Unhappy with climate change? Well, duh, vote!*

Let's try to unpack these statements. They do not directly oppose change per se. The disagreement comes in *how* that change should or could be created. Is it through voting? Is it through making people more or less angry? A social scientist would recognize that behind all these statements lies a certain "theory of change": an idea about how human societies supposedly move from one state to another, and what needs to be in place for that to happen.[8]

Scientists also operate under theories of change, like any actor in society. We believe that *in some way* what we do—our science—has an effect on the world around us. Talking about climate change, writers Rupert Read and Jem Bendell remark that one theory of change is particularly dominant: "that climate scientists slowly establish the facts, and policy makers then act on them."[9] In the follow-up to the Glasgow bridge blockade, Charlie went on a tour visiting different UK universities, trying to convince other scientists that this theory did not hold much water. In his talks, he decomposed the dominant story into at least two supplementary paths. Scientists, he argued, implicitly assume these paths allow their work to extend beyond the academy, to influence the outside world.[10] Paraphrasing him, the two paths could be stated as follows:

- Path to climate action #1: If scientists provide enough knowledge to policymakers, then they will listen to the scientists and make the right decisions.
- Path to climate action #2: Even if the policymakers in charge refuse to listen, scientists can still provide knowledge to the wider public. The public will then vote for new policymakers, who will make the right decisions.

This is just a simplification, of course (what qualifies as a "right decision"? to the benefit of whom? and which policymakers are scientists talking to?). But in broad strokes, belief in these paths lies at the core of many behaviors scientists adopt. The expectation is that, even if politicians fail to act on climate change, a more informed public will eventually vote for representatives that will.[11] So that's what we, the scientists, need to do: inform. No less, no more. This theory of change is just that—a theory—a story we think explains the facts we are seeing. In this case, though, the facts seem to point in a very different direction. As Bendell and Read recognize, "the story isn't true. It isn't working. And: we're out of time."[12]

Blocking a bridge is not a behavior that naturally emerges out of this theory of change. The scientist-activists in Glasgow were not providing any new information, neither to the public nor to the politicians meeting at the COP conference. Indeed, while prevalent in academia, this theory of change is not espoused by the SR movement. Like Bendell, Read, and Gardner, SR activists also believe this story is not working. To better understand the wave of scientific rebellion that hit the streets in 2021 then, we must question *why* this story may not be working and what other stories might explain the facts better. This, in turn, might lead us to understand the theories or beliefs scientist-activists are espousing, the ones leading them to conclude that breaking the law is worthwhile.

Let's start with that precious commodity that seems to be growing scarcer and scarcer: time. A core component of both paths in the dominant theory of change is electoral politics: a slow, meandering and highly bureaucratic process for decision making. Climate and ecological breakdown, in contrast, is fast (indeed it is already happening all around us). Even assuming politicians are informed and willing to act, politics still takes time. And the public knows this. A recent poll shows that around 84 percent of people in the European Union recognize their governments have been too slow to respond to the climate crisis.[13]

To put things into perspective, let's take the example of the European Parliament, which has recently adopted a much-touted Nature Restoration Law,[14] proposed as a way to "restore ecosystems for people, the climate and the planet."[15] The law states that EU countries must "restore at least 30% of habitats in poor condition by 2030, 60% by 2040, and 90% by 2050."[16] Like all other bills, this law had to go through significant hurdles before it could become part of the EU's legal code. An EU bill must first be drafted through the European Commission, in consultation with various stakeholders. Then the bill is forwarded to the European Parliament and to all member countries, which are asked to provide opinions. Then, the Commission and the Parliament begin a debate on the proposal, which includes the election of a special rapporteur, who must then draft a report, which is further discussed, amended, and voted on in a separate committee. This amended report is then debated in a plenary session, together with reports prepared by other committees. A parallel process occurs in the European Council, which involves input from political parties and experts from member states. After this, a second round of reporting, amendment, and voting takes place in the Parliament before the final signing can even take place. The whole process generally takes at least a year, sometimes two or even three.[17] And this is all before EU member governments have even started to put the law into practice within their own legislative domains.

Conservatively, during the time it took for the Nature Restoration bill to be turned into an EU law (about 1.5 years), approximately 100 known terrestrial vertebrate species went extinct worldwide (and that's without counting those not yet discovered).[18] It is hard to narrow down how many of these extinctions occurred in Europe, but we know that already about a fifth of vertebrate species in the continent are at risk of disappearing.[19] Once they are gone, no law can bring these species back from the dead. And many of those living in unprotected areas (the 70 percent of habitats that will remain "in poor condition" even into 2030) remain at risk of extinction for the foreseeable future. Globally, the current risk of extinction is at least 100 times higher than the background rate (the number of species that would be going extinct each year if our modern civilization were not in the picture).[20] This is why the law cannot catch up. We have developed a highly efficient way to rapidly kill everything around us—and a slow, inefficient mechanism for slowing down, let alone stopping, the mass killing.

A second problem with the dominant theory of change is the way politicians evaluate information in light of their interests. Just a few weeks after Charlie Gardner was arrested for painting Rupert Murdoch's media HQ (and a few months before joining the Glasgow bridge blockade), he and fellow members of Scientists for XR published a landmark paper challenging conventional academic behaviors. In it, they pointed out that scientists often assume they have a direct line of communication with politicians but that this is rarely the case.[21] Time and time again, politicians have ignored—and often acted against—scientific advice, even while making scientists believe they are being listened to. Dr. Gardner and his colleagues stressed that scientists are not the only ones speaking into the ears of politicians. Politicians often pay more heed to corporate lobbyists—especially when those same lobbyists are funding their election campaigns. In the year 2020 alone, fossil fuel corporations spent over $136 million in contributions to political candidates, as well as $110 million on lobbying in the United States.[22] The interests of politicians thus become better aligned with those of lobbyists than with scientific advice. Today, the actions of governments can be more clearly understood as serving the interests of these lobbyists than as serving the interests of the broader public—a phenomenon known in the political science literature as "state capture."[23] Even when scientific messaging appeals to safety or common interest, at the end of the day, money talks louder than words.

One could argue that scientists should seek to position themselves in closer contact with those who enact policy. Indeed, this is one of the reasons why there are international reporting bodies like the IPCC, as well as policy councils that are informed by these bodies, like the COP. But neither the IPCC nor the COP are free from people seeking to co-opt them to their own benefit. In 2022, just a year after the Glasgow bridge action, fossil fuel lobbyists made up a larger delegation than any nation at COP27, the 27th Conference of the Parties. Even as climate researchers were speaking in the conference's auditoriums, new deals for fossil gas extraction were being signed between state representatives and corporate executives outside the halls.[24] Then, in 2023, the 28th COP was presided over by Sultan Al Jaber—the CEO of the Abu Dhabi National Oil Company.[25] During COP28, Al Jaber famously said there is "no science" behind demands for phasing out fossil fuels—contradicting decades of IPCC reporting and hundreds of scientific papers.[26] One year later, the 29th COP took place in Azerbaijan—a country whose economy is centered around oil and gas production. That time, the COP president (a former vice president of Azerbaijan's state-owned petroleum company) called oil and gas "a gift from God" as he opened the ceremony,[27] and one of its senior officials was caught promoting new gas deals in the lead-up to the event.[28]

The IPCC—while not as egregiously corrupted as the COP—is nevertheless still subject to the influence of actors with anti-climate agendas. The scientific report is hundreds of pages long, so a "summary for policymakers" is drafted every time a new report comes out. The summary is meant to highlight the main conclusions of the longer text. However, it is also subject to revisions, by politicians from all states represented in the IPCC, before it can be published. In 2021, the summary of the third part of the IPCC's Sixth Assessment Report (addressing climate change mitigation) was leaked by a group of IPCC scientists before the country representatives had a chance to edit it. The leak, delivered anonymously to journalists via members of Scientist Rebellion, was due to fears the document would be watered down by politicians and their lobbies before it saw the light of day.[29] When the official summary came out in April 2022, these fears were shown to be founded.[30] Statements about the need to close coal and gas plants had disappeared from the summary, as had references to carbon emissions from the car, plane, and meat industries. The emphasis that had been placed in the document about the responsibility of the richest 10 percent also vanished. Instead, stronger emphasis was put on placing faith on technological solutions like carbon capture.[31]

Thus, the dialogue between scientists and policymakers (when it exists at all) does not occur in a void, but is heavily affected by political and corporate actors who have agendas in conflict with scientific advice. These actors promote misinformation, often in the form of well-funded campaigns.[32] The failure of scientists is not a failure at delivering enough information or at doing it in a clear enough way. The failure lies in assuming information delivery is occurring in a void, unhindered by other interests.

Corporate actors are also interfering with scientific messaging toward the broader public—obstructing the second path to climate action in the dominant theory of change. For example, the fossil fuel industry has known for decades that rising emissions would lead to climate breakdown. Exxon scientists ran climate projection models in the late 1970s and found that its oil and gas products could lead to "dramatic environmental effects before the year 2050," and recent investigations have revealed that TotalEnergies and Shell knew about the links between fossil fuels and global warming since at least the 1970s and 1980s, respectively.[33] The industry's reaction to this information was not to change course in light of it, but to hide it from the public. Using Big Tobacco's handbook, fossil corporations have spent millions of dollars each year to occlude the evidence and amplify misinformation about climate change.[34]

Since the early 2000s, as climate change became harder to deny, corporate strategies have shifted toward making the public believe that the phaseout of fossil fuels is either impossible or unnecessary. Recently, sustainability scientist William Lamb and colleagues classified "discourses of climate delay" into different categories.[35] These include arguments in favor of surrendering to climate change ("change is impossible"; "anything we do now is too little or too late"), efforts at redirecting responsibility ("what about China?"; "what about your own carbon footprint?"), or efforts at emphasizing the (often false) disadvantages of renewable energy ("solar and wind are too expensive"; "we need fossil fuels to get people out of poverty").

Scientists have sometimes responded to misinformation campaigns, but usually in dispassionate ways and almost always by asking nicely of those in power: those with an interest in maintaining the status quo. Instead, as IPCC author and social ecologist Dr. Julia Steinberger argues, scientists should be talking "about power" rather than "to power," by engaging with mass protest movements:

I believe the best antidote to these massive machines of disinformation, which are still at work and will continue to wreak havoc all around us, is to get out there and be personal, passionate, insistent voices for change. We can have integrity and activism at the same time: indeed, how can we be true to the implications of our scientific findings if we do not?[36]

Unlike other IPCC authors, Steinberger not only talks the talk, she walks the walk: she is a member of the climate movement Renovate Switzerland and went viral when she joined a highway blockade in the Swiss canton of Bern. As she was being arrested and escorted into a police van, she remained stoic: politely relaying the need for urgent climate action.[37]

HEGEMONIC DISCOURSES

Now that we are back to talking about power, we can tackle one last weakness of the dominant theory of change. To me, this weakness is the final nail in the coffin, perhaps the most important reason why the dominant story is not really working; why relying on conventional forms of scientific outreach and expecting politics to react to them is simply not enough. To talk about this weakness, though, we need to first introduce the concept of cultural hegemony. Originally developed by Italian intellectual Antonio Gramsci, the theory of hegemony permeated much of Marxist and post-Marxist thinking in the second half of the twentieth century. It was introduced in Gramsci's *Prison Notebooks*, written while he was in jail at the height of Mussolini's fascist regime.[38] Gramsci explained that power can be maintained via the threat of physical coercion: someone with a military, police, or private security force is more likely to be obeyed than someone without it. But power can also be maintained through cultural hegemony: spreading those values and norms favored by the elite, while ridiculing or repressing those who run counter to their interests. Gramsci states that any hegemony is fueled by a "common sense": a set of values and norms that the elites favor and that are made socially acceptable. This, in turn, prevents those ideas that could challenge the dominant class from taking hold and becoming widespread. Just like Dr. Pangloss in Voltaire's novel *Candide*, those in power perpetuate the notion that we live in "the best of all possible worlds."[39] The easiest way for them to do so is through instruments of mass cultural diffusion, like the press, education, and cultural institutions. Common sense thus manufactures widespread consent in favor of the prevailing social order.[40]

In a recent book, social scientists Daniel Nyberg, Christopher Wright, and Vanessa Bowden explain that to understand climate inertia—the lack of real political will to curb emissions—we need to see it through the lens of Gramsci's hegemony.[41] Fossil fuel companies often advertise themselves as being key actors in the "green transition" or as the only path to "affordable energy." They do this even as they keep pumping billions of dollars into new oil wells and coal mines and lobby against cheaper, renewable forms of energy. Thus, fossil capitalists and their partners in government establish artificial equivalences between their goals and socially desirable ideas. This helps make them less likely to be attacked by the public—they are with them after all; they are the "good guys."

This is also the reason why fossil fuel companies often seek to sponsor famous museums and art galleries: in doing so, they associate their brand with societal actors that already enjoy high levels of respect or support. A prime example is the British oil company BP (formerly known as British Petroleum), which has sponsored exhibitions at the British Museum for over 25 years.[42] This has allowed it to slowly seep into the cultural ethos of the British public, effectively using the museum's prestige to launder its own image.[43] Closer to the world of science, another major institution in bed with fossil capital is the London Science Museum. The museum has worked with BP and the energy company Equinor on the creation of entire galleries on energy and fossil fuels and has hosted exhibitions codesigned with oil giant Shell on carbon capture and storage.[44] In 2015, leaked internal emails revealed how Shell was even trying to influence the museum's climate program.[45] Most recently, the Indian coal conglomerate Adani has also joined the fossil sponsorship party, providing funding to the museum's new exhibition, named Energy Revolution: The Adani Green Energy Gallery.[46]

A constant tug of war exists between the cultural hegemony and the actors whose interests are in opposition to it.[47] The relationship between the hegemony and these other actors is never fixed, but constantly contested. Often, when challenged by external actors, fossil capital has welcomed some external actors into its hegemony, changing the nature of its discourse in the process. For example, fossil fuel companies might partner with green NGOs in campaigns seeking to improve their image, or fund "green projects" in universities to push their brand, while still engaging in highly polluting activities. Alternatively, the hegemony might exclude more-challenging movements, creating a polarized narrative, so as to frame them as *the common enemy*. In a November 2022 televised speech, UK home secretary and Conservative Party member Suella Braverman said that climate protests are "a disruption to our way of life." She urged police to take all measures necessary to stop climate activists, equating them with extremist vandals: "I fully back police in using all tools available to prevent further disruption and protect the public."[48] Similarly, in September 2023, the Spanish national prosecution office went one step further and decided to classify climate movements that embrace civil disobedience as "terrorist groups."[49]

These types of narratives are examples of hegemonic discourses: "our way of life" implies that the welfare of society and of fossil fuel corporations are one and the same. They are wrapped together with fossil capital into a common *We*. The rest of society is told to not question this articulation. The call to "protect the public" then serves to draw a line in the sand—to exclude climate activists from the universalized We. They are the Others, the *terrorists*. Thus, activists are portrayed as targeting not fossil fuel corporations, but rather the public in general: *you and me, and our way of life.*

Until recently, scientists have tended to shy away from critiquing (or even questioning) these hegemonic discourses. In fact, scientific communication tends to almost operate as if connections between the pillars of fossil capitalism either don't exist or are just not worth talking about. By seeking to stay "apolitical," scientists have let existing articulations—the fossil hegemony—go unchallenged.

BUILDING COUNTER-HEGEMONY

How can the pairing between fossil capital and the rest of society be dislodged? The political theorists Ernesto Laclau and Chantal Mouffe[50] argue that we can build an alternative narrative that can mobilize against existing power structures—a so-called counter-hegemony.[51] For example, Nyberg, Wright, and Bowden propose building counter-hegemony by externalizing fossil fuel corporations (and their economic, social, and political pillars of support);[52] in other words, we can label them as the new Other—opposed, in their interests, to both climate movements and the rest of society. This can help steer us away from discourses in which capital is seen as a partner rather than the enemy.

Along these lines, the organization Culture Unstained[53] has carried out a series of campaigns over the past two decades, to drive fossil capital out of various British cultural institutions.[54] They have used a wide array of creative tactics designed to shame museum boards for having fossil partnerships. These tactics have included sneaking a giant wooden horse into the British Museum to symbolize that—like the Trojan horse of ancient Greek lore—BP's sponsorship of the museum is a deadly menace in disguise.[55] A decade after its inception, the organization has met with enormous success. They managed to get the National Portrait Gallery, the Royal Shakespeare Company, and the Tate museum to end sponsorship deals they had made with BP.[56] Campaigns in these institutions have made the social cost of associations with the oil company higher than the financial benefit that could be gained from them.

Similarly, the London Science Museum has been subject to actions designed to break its many fossil sponsorship deals. In this case, scientist-activists have played a crucial role: since its inception, the collective Scientists for XR has worked with Culture Unstained to carry out protests both outside and inside the museum,[57] interrupting panel debates[58] and even going as far as camping inside Adani's Energy Revolution gallery, delivering public lectures, and building sculptures made of lumps of coal.[59] The Science Museum's director— Ian Blatchford—insists on maintaining the fossil sponsorships, but the campaign has had a lasting impact on his image and that of the institution.[60] A long list of scientists, teachers, and former exhibit organizers have announced that they will no longer work with or pay visits to the museum, at least until these partnerships end.[61] Public events and exhibits inside the museum have had to be canceled after speakers and gallery contributors chose to withdraw.[62] And under mounting public pressure, two members have now resigned from its board of trustees.[63]

Thus, scientist-activists can be important players in the creation of counter-hegemony. We saw earlier that scientists are trusted members of society.[64] We also saw that most science communication today operates as if there is a direct line of conversation between scientists and policymakers. Indeed, the core mission of any university's public relations office is to highlight how much their research has "impact" and "real-world applications," how much they can help inform society's leaders and improve our lives.[65] Scientists also operate as if politicians are the actual agents of change—those working for the interests of the public, and those who need to be informed. In the language of Gramsci, all these "as ifs" are "relations of equivalence" by which scientists and politicians are brought together under the same articulated discourses. But these "as ifs" can also be contested. Scientist-activists can build counter-hegemony by exposing these relations of equivalence between scientists and politicians as fundamentally flawed—and in need of rethinking.

Nyberg, Wright, and Bowden stress that a way to break existing relations of equivalence is by expanding the realm of the "political": those things in our societies that are potentially subject to change.[66] After all, *our representatives in government have known what to do for decades, but they have refused to do it.* This, then, points to a fundamental flaw in the way scientific advice "flows into" change, with politicians as middlemen between scientists and the rest of society. The key then is to make the public aware of this flaw and of the consequences it has for them—to break the illusion that politicians are actually acting on the science.

PUSHING THE POLITICAL FRONTIER

This brings us back to the first wave of actions by Scientist Rebellion. Mike's action at Downing Street wouldn't have been necessary *if* the prime minister was in fact acting according to scientific recommendations. His action shined a light on the flaw of that assumption—the tacit connection between science and politics was momentarily exposed and ruptured, for the rest of the public to see. The pasting of scientific papers on the walls of government ministries is also an exposure: these acts make it obvious that politicians are not listening to scientists, once the scientists stop acting *as if* they are. The bridge blockade at COP26 was an exposure of failure at a more global level: a rupture of the connection between science and the intergovernmental body supposed to be tackling climate change mitigation. Dr. Gardner explained his reasons for joining the action: "This COP has failed, just like the previous 25 COPs have failed, and we cannot rely on our leaders to do their job anymore."[67]

All of these actions involved scientist-activists questioning an articulation of the fossil hegemony—showing that our common sense may not be so sensible after all. They also served to build new bridges, new "relations of equivalence": this time not with politicians, but with the public at large. In their earnestness and vulnerability, these scientist-activists served to make scientists look more human, to show that they too fear for the future. They also pushed the limits of the political—separating what can and cannot be contested.

Ultimately, scientist-activists were challenging hegemonic limits to established scientific behaviors; they were calling into question whether what we often think a scientist *should* be doing (running experiments, writing papers, giving talks) is what they should *actually* be doing. They evinced that scientific behaviors always have a political connotation. Be that through their action or inaction, scientists have an influence over how the public thinks about scientific advice and the government's response to it. When the scientists—the experts—start behaving as if our leaders are not listening to them, that's a wake-up call for the public, an alert that something is off, that science and policy are not working together after all.

CHAPTER 3

Education as Resistance

"The street was the classroom; the bike lane was the blackboard."

—University Post, Copenhagen, October 26, 2021[1]

The spirit of scientist disobedience eventually reached the small country of Denmark, where I work. The first protest by Scientist Rebellion here was a seminar. It did not take place in a lecture hall, though. The venue was the street in front of the Ministry for Climate, Energy and Utilities, in central Copenhagen.

It was October 2021. A recent election had given an overwhelming "green" mandate to the Danish government. In the following months, a coalition of center-left parties had passed the first Climate Act in the history of the country.[2] It seemed like things were finally starting to move; for example, the Danish government was finally willing to drastically curb the country's emissions. The law dictated a path that present and future governments would have to follow in order to reduce 70 percent of Denmark's CO_2 output by 2030. It was not a fully satisfactory goal—for example, it neglected to include Denmark's overseas and transport emissions—but it was something measurable, something the government could be held accountable for.

As the months passed, the government began reneging on their promises. The climate minister at the time—Dan Jørgensen—proposed a "hockey stick" plan for meeting the requirements of the Climate Act: do very little now and a lot in the last few years (the blade of the hockey stick), right before 2030. This plan allowed the present government to pass the onus of decarbonizing on to the future, to those who would be in power many years from then. It also involved focusing on investments in undeveloped technologies, like carbon capture, instead of reducing emissions in the here and now.

At the time, I was meeting frequently with fellow scientists who had been active in the Fridays for Future marches of 2018. We were frustrated, and we were angry. We started brainstorming ways to show that Dan Jørgensen's plan was not what the academic community meant when talking about "urgent climate action." As scientists, we wanted to show the public that this plan ran against IPCC recommendations and would lead the government to effectively break their own law. One of us—political economist Dr. Laura Horn—came up with a proposal: "Why don't we teach a class at the ministry?"

It was a brilliant idea. Deliver the facts of the climate emergency at the doorstep of those in power. If they refused to listen to the science, we would bring the science to them. It was not so much about teaching *them*, of course, but about building a counter-hegemonic discourse, about showing the public that scientists were saying one thing and politicians were doing something completely different. Like the Glasgow bridge blockaders, we wanted to break the relation of equivalence between scientific advice and political will.

We invited 11 professors and academics who were knowledgeable about climate and ecological breakdown to each give a 10-minute lecture on their particular subject of expertise. We included ecologists, sustainability scientists, political scientists, and sociologists. We invited the public to attend the lectures as well. People could listen to the professors from the sidewalk, but the idea was for the classroom to extend into the street as well; those who wanted could choose to sit on folding chairs placed in the middle of the road. Thus, people could be part of the audience while blocking traffic.

Early in the morning of October 26, we gathered in front of the ministry. When the light went red and traffic stopped, we moved quickly. We placed a fruit crate on the bike lane next to the street and used it as a podium on which the lecturers could stand. Each of us also brought a folding chair and quickly placed it on the street. To each side of the "classroom," activists and scientists were standing and holding banners showcasing our message: *listen to the science* (chiefly what politicians were refusing to do). The banner holders were also there to protect the audience and indicate to the drivers that the street was blocked on both sides. Once everything was in place, the seminar began.

I was tasked with being master of ceremonies, together with Aitz-koa Lopez—a fellow scientist-activist working on medical genetics at Lund University in Sweden. Hesitantly at first, we stepped together onto the podium, grabbed a microphone, and opened the event. We welcomed all who were in attendance—professors, students, activists, and regular passersby, curious to know what the hell was going on. We also turned around and gave a warm welcome to Dan—the climate minister—knowing he could be watching from somewhere up above in his office. Finally, we introduced the first lecturer. The seminar was now officially under way.

The police arrived soon after the first talk had started. It was a bizarre image. Students calmly listening to a professor giving her lecture, surrounded by police officers who had to divert car and truck drivers. The event had been announced in advance (though not its exact format), and there were a lot of journalists on-site. We were interviewed for TV, many of us for the first time. I remember being very nervous. This was the first time I had ever broken the law. And I was doing it on national TV. I was used to organizing lectures; I had done it countless times as an academic. Yet I was not used to a seminar that could conclude with me in handcuffs. It was a first for the Danish police as well. They were accustomed to moving activists off the road, but they did not know what to do with academics wearing lab coats, calmly giving a lecture to a class of students. With each new talk, the police chief was getting more and more irritated with us. At one point, he told us: "Either you move or we move you."

As I was listening to the talks, I became more relaxed. It was a sunny, warm day (an unusual blessing for Copenhagen). And there was something calming, even ethereal, about attending a lecture in such an unusual space. Eventually, I was able to forget about the police and pay attention to the speakers. Their lectures were fantastic. Dr. Laura Horn was the first to speak, and she made full use of the space we were occupying. She was speaking on a bike lane, slightly elevated above the rest of the street (bikers, unlike cars, were allowed to go through). In the middle of her lecture, she took out some chalk and started using the lane as a blackboard. With colorful graphs drawn directly on the pavement, she explained how most countries had not yet decoupled their carbon emissions from economic growth and that those that appeared to be successfully decoupling, like Denmark, were doing it far too slowly to meet the limits of the Paris Agreement.

The rest of the lectures were about related topics: about the countless species being driven to extinction, about the misleading narratives corporations use to get us to forget about their pollution. They were also about the roots of climate breakdown and the history of imperialism and capitalism, about the many connections between environmental struggles and feminist liberation, about deliberative democracy and its potential for helping stir a radical transformation of our societies so they could be back in equilibrium with the rest of nature.

After a while, the situation de-escalated. We had a well-trained activist engaging in dialogue with the police, and he helped to calm things down. At the time, the police agents did not want to be seen arresting academics. After a few phone calls with central headquarters, the police chief told us they were declaring our blockade "legal" and that we could continue with our roster of speakers until the end, as long as we left after the last lecturer had finished their talk.[3] We held our full seminar for its planned duration of 3 hours. During that stretch of time, the street by the ministry was not a place for cars; it was a place for collective learning.

The event was covered by every major news outlet in Denmark—not just what one could call progressive media,[4] but also mainstream

FIGURE 8. Dr. Laura Horn speaking during the teach-in at the climate ministry in Copenhagen. Credit: Fernando Racimo.

public and corporate newspapers and TV channels,[5] academic gazettes,[6] and even a right-wing tabloid.[7] Late into the evening, we kept getting calls from journalists across the country, asking us to retell what had happened.

RELOCATING THE CLASSROOM

Our protest was an example of a teach-in: a lecture or discussion held to address a perceived injustice in society, outside the regular curriculum. Our intention was to create something that was academic but still out of the ordinary, *out of place*. This displacement of teaching into the street was a central focus of our action. Frederik Appel Olsen—a PhD student in rhetoric—was one of the academics who helped organize the event. He could not be present at the time (he was on a student exchange abroad), but after watching the events unfold on Danish media, he decided to make the teach-in a central topic of his doctoral thesis.[8] In it, he explains why the event was so effective at garnering public attention:

> In Scientist Rebellion's teach-in, place . . . provides an important rhetorical function. This is not simply due to the fact that social movements' protests and demonstrations as a rule unfold in carefully chosen places in public space to create visibility and attention, but also to a large degree to the fact that it is the movement from a usual and expected place (the scientific institution) to an unusual and surprising place (the street in front of a ministry) itself that functions as a point of attention and a source for questions about scientific ethos in Scientist Rebellion's action. . . . The material relocation of the classroom is as central in the media coverage as the question of the researcher/activist role.[9]

This displacement of teaching into the street serves to raise attention to the underlying conflict at the heart of the protest. Scientists step out of the comfort of the classroom as a way to show that those in power are ignoring a crisis that affects us all and that politicians urgently need to act on it.

A teach-in also puts those in power in an uncomfortable position, and it alerts society to the inconsistencies between words and deeds. On the day of the protest, the climate minister sent a reply to the journalists who insistently called his office, asking what he had to say about the teach-in. While trying to defend himself, he said Denmark had "probably the world's most ambitious policy in this area."[10] This was hard to reconcile with the events going on at his doorstep. If scientists were risking jail to say his climate policies were falling short of what was needed, then how could he claim these were ambitious? Politicians often speak as if they are acting according to scientific findings. The teach-in at the ministry stripped away this supposition and laid bare the ugly truth: most of the time, it is not science that politicians base their actions on. Other, conflicting interests are at play, preventing action, and the public needs to know about them too.

This was not the only time that scientist-activists displaced their classrooms as a form of protest. Between 2021 and 2023, climate teach-ins were held inside university lobbies, at corporate headquarters, and on city streets on all four corners of the planet: in Sweden, Malawi, Colombia, Nigeria, Rwanda, and Argentina.[11] Many of these teach-ins involved academic speakers delivering seminars together with people who have been directly affected by extreme weather events—who have seen firsthand how fires or floods wreaked havoc on their own communities. Here, the displacement of the classroom has an even more powerful symbolism: it echoes the (forced) displacement of people from their own homes.

And, sometimes, the displacement of the classroom can be a way to reclaim an area. For example, 2 days after their bridge blockade in Glasgow, Scientist Rebellion activists managed to hold a teach-in at the same location where they had been arrested. Charlie Gardner told me:

> After we left the police station, we were reading the bail conditions under which the police had released us. They had banned us from the entire area where COP was taking place, but we realized the ban only extended up to the limits of the Glasgow jurisdiction. The conditions did not forbid us from going back to the same bridge we had been arrested at, because half of the bridge was actually outside of the Glasgow jurisdiction! Well, we thought, if the police are not going to exclude us from our own bridge, then we're going to go back there. So, 48 hours later, we did a teach-in at the same bridge, delivering lectures about the climate emergency to the public.

A NIGHT AT THE MUSEUM

Given the importance of *place* in the teach-in, scientist-activists often look for iconic locations to hold their disobedient classes. A few months after our action in Copenhagen, on the night of April 9, 2022, 30 scientists held a seminar while occupying the Muséum National d'Histoire Naturelle—a hallmark institution of natural history in the French capital. The scientist-occupiers—among them ecologists, marine biologists, and historians of science—gave 12 presentations to call for urgent, radical climate policy in France and abroad.[12]

Surrounded by fossils of species long gone from the face of the planet, their action was an allegory of the threat our own species faces today: *we too can become fossils if we don't act now.* During the seminar, several of the scientist-activists were even locked to the feet of a mammoth skeleton—the centerpiece of the hall.[13] As one of the participants, Thierry de Novadhau, explained to me:

> We selected that hall because it had large fossil skeletons, and so it helped us to illustrate the sixth mass extinction happening today. We thought presenting some scientific facts about extinction from different domains and subjects was something that could work as a form of outreach to the public.

During an intermission between two of the scientific talks, one of the seminar participants read a quote by the late biologist Rachel Carson, evoking the feeling of extinction closing in on her, as she imagines walking on a pesticide-polluted landscape:[14]

> There was a strange stillness. The birds, for example—where had they gone? Many people spoke of them, puzzled and disturbed. The feeding stations in the backyards were deserted. The few birds seen anywhere were moribund; they trembled violently and could not fly. It was a spring without voices. On the mornings that had once throbbed with the dawn chorus of robins, catbirds, doves, jays, wrens, and scores of other bird voices there was now no sound; only silence lay over the fields and woods and marsh.[15]

Like the museum occupiers, Dr. Carson was both a scientist and an activist. During the 1950s and 60s, she fought fiercely against powerful polluting corporations, paving the way for landmark legislation against toxic chemical dumping in the United States. She also helped other scientists realize they could be more than knowledge gatherers, that it was their duty to participate in the policy debates of their time.[16]

Here too, the displacement of the classroom served to expose the underlying contradictions at the heart of French politics. The museum occupation took place in the middle of the national electoral season, and it followed after a letter signed by over 1,000 scientists published

FIGURE 9. The SR teach-in inside the Muséum National d'Histoire Naturelle. Credit: Scientifiques en rébellion.

in *Le Monde*.[17] The letter called for mass civil disobedience in the face of the constant inaction of the government in France to curb its emissions. It also demanded a transition away from industrial agriculture, toward ecological land use practices.

The scientist-activists managed to stay until midnight inside the museum, streaming their lectures online long after closing hours. They were practicing what the letter to *Le Monde* preached, going from mere words into action. They were also honoring the work of Dr. Carson and the many activists who helped jump-start the global environmental movement back in the 60s. During the occupation, a large banner hung upon the museum's inner balcony reading "Dire la vérité n'est pas un crime" (To tell the truth is not a crime). Thierry described to me the reaction they received from those inside the museum:

> We came just before closing time because we didn't want to perturb the public, but we wanted to stay there after it closed. Everything went really smoothly at the beginning. The security staff was initially annoyed because it meant that they had to stay there because of us, but at the end they did

thank us because they had learned a lot. The museum director was not happy. He was the one that called the police. Many of the scientists working inside the museum were really upset when they later found out the director was pressing charges on us. After the action, they organized a support campaign for us, saying that of course they didn't agree with the director.

At the time of writing, eight of the scientists have faced trial before the Paris police court, and five more are expected to do so in coming months.[18] As in Denmark, the teach-in and subsequent trials have helped put the French chapter of Scientist Rebellion on the map, amplifying the activists' message well beyond the museum's halls. SR membership in France now numbers in the hundreds.

CLIMBING BACK UP THE SHOULDERS OF GIANTS

Teach-ins were not invented by Scientist Rebellion. They have been at the heart of struggles for justice for decades. Our intervention at the ministry was not even the first teach-in for climate action to take place in Copenhagen. Just over a decade before, faced with the policy failures of the 2009 COP15 conference in Denmark, members of the Climate Justice Action collective chose this mode of protest as well. Their "academic seminar blockade" took place while they blocked the gates of a coal-fired power station near the Copenhagen harbor. As seminar organizer Kelvin Mason tells it, the appearance of the lecturers also deterred the Danish police from interrupting the event. In this case, the academic symbolism emanated from "reclaimed" conference name tags instead of lab coats:

> The police eventually allowed our action to proceed unhindered, seemingly because of the privileged status they conferred on us as academics. This status was conveyed to them principally via the name-tags we wore, even though these were reclaimed from previous academic events and bore a random selection of names (with apologies to any reader who is subsequently subject to arrest for no apparent reason!).[19]

Going back even further in time, teach-ins had an especially pervasive presence in US universities in the 1960s and 70s. Academics would give lectures with an explicitly political stance in occupied classrooms, hallways, and other unconventional campus spaces. As it does today, the displacement of teaching occurred because they perceived their ruling institutions were not living up to the critical issues of their time. This included the struggles for peace, nuclear disarmament, women's rights, and civil rights for people of color.

The Vietnam War was the catalyst for the first teach-in, which gave the practice its name. It took place at the University of Michigan at Ann Arbor, in early March 1965. Faced with decreasing troop morale, and the realization that they were losing the war, the US government had just launched Operation Rolling Thunder: a sustained carpet-bombing campaign that would last until 1968. The campaign would cover the Vietnamese countryside with over 800,000 metric tons of bombs (almost twice the amount deployed against Japan during World War II) and claim the lives of over 30,000 civilians.[20] In many parts of the country, entire villages turned instantly into scorched wasteland.

From its start, the campaign instigated major backlash in the United States, and universities were the epicenters of resistance. In Ann Arbor, soon after the start of Rolling Thunder, around 50 faculty members tried to go on an anti-war strike. However, they faced repression from both the university president and the Michigan state government, who called their strike illegal. The professors feared for their jobs, but they were willing to stand their ground. Anthropologist Marshall Sahlins came up with an alternative: "They say we're neglecting our responsibilities as teachers. Let's show them how responsible we feel. Instead of teaching out, we'll teach in—all night."[21]

Sahlins and other academics carried out the first teach-in in a large hall inside the university, on March 24 and 25, 1965. It was attended by over 3,000 people, occupying the campus overnight. It included lectures, debates, and artistic performances, all in protest of the war. Inspired by the famous sit-ins of the civil rights movement, this type of protest added an intellectual element to the civil resistance arsenal. As Sahlins puts it, "it was a near-perfect synthesis of being in the system and out of it, of academic responsibility and civic dissent."[22]

But like all protest that threatens power, it was not without an element of risk. The lecturers received bomb threats three times during the night. Forced to empty the occupied hall until the threat was clear, they continued lecturing out in the cold of the campus esplanade.[23] Here again, the classroom defied spatial boundaries. It happened to be wherever the lecturers and the protesters happened to be.

Soon, teach-ins were rolling out all across the country. Two months after Michigan, lecturers in UC Berkeley upped the ante with a teach-in attended by over 30,000 people (our small intervention at the climate ministry pales in comparison). Over 120 teach-ins took place across American campuses in 1965 alone, and they would continue until the end of the war.[24] Along with mass protests and strikes, the teach-ins would help turn the tide of public opinion in favor of peace.

Faced with resistance from government officials and university managers, many more groups of scientists came together in the 60s and 70s to support each other in joint struggle. The Cold War arms race and the proliferation of nuclear weapons, for example, propelled many to recognize that science was not an unqualified good to society; rather, scientific activities could be used to support and expand the war machine that they (in principle) opposed. Thus, as in the climate emergency, scientists also realized they had a special responsibility to rise up. A day of academic strikes and teach-ins at MIT on March 4, 1969, led to the formation of Union of Concerned Scientists,[25] an organization that, in its founding document, argues "for turning research applications away from the present emphasis on military technology toward the solution of pressing environmental and social problems."[26] The Union of Concerned Scientists is still active to this day, and its work now extends to bringing public attention to climate injustices. The extraction and combustion of fossil fuels is, after all, made possible by science and technology, just like the Bomb.

Another movement that emerged in the civil rights era was the Science for the People collective. Composed of progressive academics who pooled resources to mobilize for protests against war and racial segregation, the group organized not just teach-ins, but also mass marches and strikes in universities and research institutes across the United States.[27] Science for the People members held the belief that science can never be neutral. They saw that calls for academics to "stay in their lane" and remain silent in the face of social injustices only served to perpetuate those injustices. Prominent scientists of the time, like Stephen Jay Gould and Richard Lewontin, were leading voices of activism within the organization.[28] In recent years, the collective has enjoyed a revival. They helped organize the 2017 March for Science in Washington, DC, in protest against the first Trump administration's hostility to science and academic institutions.[29] The idea of carrying out a March for Science spread to over 600 cities across the world and, years later, would be a source of inspiration for Scientist Rebellion's characteristic use of the lab coat while protesting.

EDUCATING FOR THE FUTURE

Beyond teach-ins, scientist-activists are also staging smaller educational rebellions inside the classroom: interrupting their own lectures to deliver talks about socio-ecological breakdown instead of their assigned topic for the day.[30] In this case, it is not the class that is displaced, but the curricular content expected to be delivered. As I and other members of Scientist Rebellion have argued, both teach-ins and in-class rebellions are forms of "expanded academic practices" in the climate emergency.[31] Academics expand their pedagogical repertoire—sometimes beyond the limits of the legal—and embody the idea that "there is no such thing as a neutral educational process," as Richard Shaull put it in his preface to Paulo Freire's *Pedagogy of the Oppressed*, or in Freire's own words, that education must be a path for people to "deal critically and creatively with reality and discover how to participate in the transformation of their world."[32] Teach-ins and in-class rebellions are also a way to make the classroom a site for politics, a space where the political frontier can be pushed.[33] They make something contestable that previously wasn't, be that the spatial limits of the place of teaching, the content supposed to be delivered within it, or the form of its delivery. Educational rebellions remind us that, as Black feminist scholar and activist bell hooks put it, "the classroom, with all its limitations, remains a location of possibility."[34]

Dr. Julia Steinberger argues that a rebellion inside the classroom can be as simple as listening to what students have to say, what they need from their teachers. Students are, after all, the first who managed to raise global awareness about climate breakdown, at a scale that was thought impossible before. They do not need to be told how bad things are over and over again. As Steinberger herself experienced when teaching, rather than more knowledge about the *consequences* of climate breakdown, students are in dire need of tools for understanding its *causative agents*, the logics and structures of power at the heart of it. And, perhaps most importantly, they are in need of tools for *confronting* them—tools they are seldom obtaining from their lecturers.[35] In the climate emergency, the most useful education is the stuff they don't teach you at school (and that is exactly what needs to change).

So what's next? The conundrum educators face today is that our past will look very different from our future—in a way that is unlike anything faced by previous generations of educators. In a recent publication, scientist-activist Dr. Aaron Thierry and colleagues stress that

> we are training students for a future that will not come to pass . . . under what are projected to be drastically altered circumstances. Currently, we are striving to achieve professional success, but not collective survival.[36]

To attain collective survival, today's students will have to understand how to face a world not just politically or economically different from the one educators grew up in, but also biophysically transformed. Some features of this world educators will get to see in their lifetimes. Some (perhaps the most gruesome ones) they might not. We can use the budding present as a guide for what is to come. New forms of education—education *for* the future—are already emerging in different corners of academia, struggling to be born even as older, established structures fight to maintain their foothold.

Power is also at the center of today's teaching conundrum. To many, education is a process of empowerment. We transfer our powers from ourselves to others. This includes the power to understand existing ideas and systems of thought, the power to connect them and confront them, and the power to create from them anew. We hope these powers will be of use, even long after we are gone. As writer Toni Morrison used to tell her own students, "when you get these jobs that you have been so brilliantly trained for, just remember that your real job is that if you are free, you need to free somebody else. If you have some power, then your job is to empower somebody else."[37]

Chief among the powers shared in the process of teaching is the power to challenge hegemonic ideas, including ideas about how societies should be organized, about who should hold power and who should not, and about how to live a good life—ideas that were forced upon the minds of Global South peoples through centuries of domination, in what has been termed the "cognitive empire" of the Global North. Today, these same ideas continue to perpetuate ecological injustice, but they are increasingly challenged.[38] An education for the future should be an education that supports and helps strengthen the challengers, an education that confronts power and breeds social and ecological justice.

Teaching others to confront power is not something most academics are expected or encouraged to do. Power seeks to perpetuate itself and to remove the tools that can be used to confront it. Among these tools is knowledge about the history of activism, about teach-ins and educational rebellions of the past. Dr. Steinberger explains:

> The history of how [confronting power] can be done has been erased from the classroom. We don't learn how to be activists, advocates, muckraking journalists or revolutionaries at school. But we can and need to learn this now.[39]

Education is also about honoring those who empowered us in the past, by empowering new generations for the future. Here too, educators have a lot in common with activists—they are both involved in intergenerational projects. "The hardest thing to accept is that you're not going to be the one who crosses the finish line. It's not a sprint. It's not a marathon. It's a relay." Activist carla joy bregman uses these words to describe the temporal nature of social justice struggles.[40] They apply equally well to the struggle for climate justice and to the educational sites through which these struggles unfold.

Yet, a truly intergenerational, power-confronting education is not easy. Institutions today continue to place barriers on what educators can and cannot do—on what is allowed to be taught and on *how* it should be taught. In his book *Educating for the Anthropocene*, Peter Sutoris talks about how academia today dehumanizes both students and teachers. Through increasing bureaucratization, they are made to think of themselves as "cogs in the machinery"—a figure of speech denounced by Hannah Arendt for its use by members of totalitarian regimes to justify their atrocities.[41] To break out of bureaucratization, Sutoris argues for an education that not only fosters intergenerational dialogue, but also teaches how to communicate with others as political beings; as beings who are able to create, together, new ways of living with others; as beings who see themselves as more than just cogs.[42]

Sutoris also highlights "radical imagination" as a key ingredient for an engaged education. Educators need to promote in the students the power to think new futures, new stories of what is to come, before they can be made real. Unlocking radical imagination will involve a great degree of "unlearning"—of undoing and unraveling ideological bonds that remove our agency in the construction of new futures and allow the wheels of capital to keep on churning.

These bonds include those that tie academia together with fossil fuel corporations and other destructive industries, but also bonds that are internal to academia itself. They include the mental bonds that keep both students and teachers constrained to individual disciplines, and to hierarchies of power within those disciplines, and that restrict us to limited action within closed-off departmental silos. As climate justice scholar Jennie Stephens puts it, we have, in many ways, been taught to "be like fish swimming around in a fishbowl."[43] Educators for the future will need to help students to "swim in the ocean," by breaking down cognitive constraints that may not even be visible to them today.

PREGUNTANDO CAMINAMOS

I would like to add one more ingredient to Sutoris's ecopedagogic mix: *communal courage*. In order to break existing barriers to action and help create radical new futures, educators will need the courage to defy established rules—those rules that limit what teaching can look like—in a world that will look like nothing that has come before. But we cannot do this in isolation. Courage needs to be nourished by community—by supporting and being supported by others who are also choosing to break the rules.

That October morning at the ministry teach-in, as I was stepping into the street, taking my baby steps into civil resistance, I was not doing it alone. I was together with Laura, with Aitzkoa, and with many other academics who were choosing to be much more engaged than their institutions allowed them to be. We were also in dialogue with the giants of the past, with Rachel Carson, with Kelvin Mason, with Marshall Sahlins, and with the countercultural rebels from Michigan and Berkeley—those who took these same steps years before some of us were even born.

As scientist-activists, we are often walking into the unknown, not knowing what the repercussions of our actions will be or what our fellow colleagues will say about us once they see us breaking the rules. We borrow on the Zapatista motto—"preguntando caminamos" (asking, we walk)—a call to embrace radical uncertainty, to create the path as we walk it ourselves.[44] To me, the greatest wisdom of this motto lies in its plural form. *We* walk not knowing what lies ahead. But we never walk alone.

CHAPTER 4

Out of the Lab and into the Streets

"What's the use of having developed a science well enough to make predictions if, in the end, all we're willing to do is stand around and wait for them to come true?"

—Frank Sherwood Rowland[1]

In December 2022, the American Geophysical Union (AGU) was hosting its annual conference. AGU is the largest society of climatologists and geologists in the world. Hundreds of attendees were present, discussing their research findings and delivering talks to their colleagues. In the middle of one of the conference's plenary events (the largest of all the talks), Earth scientist Dr. Rose Abramoff and climate scientist Dr. Peter Kalmus stood up from their seats in the audience and did something unexpected: they rushed onto the main stage, unrolling a protest banner.[2] In large font, for everyone all the way in the back to read, the banner contained a single sentence: "Out of the Lab & Into the Streets."

Dr. Abramoff and Dr. Kalmus—both SR members—were calling on their colleagues to step out of the comfort of the ivory tower. They were arguing that Earth scientists were not being loud enough about how rapidly the Earth is destabilizing. Dr. Abramoff explained to me why they chose to interrupt the session:

> We did not want to attend another conference without somehow pushing against the status quo. It felt to us that just interacting with all these sci-

entists and not motivating them to take action did not seem to be a good use of our time. We were trying to snap scientists out of their regular programming.

Though the protest lasted 32 seconds and was well received by the audience, the conference organizers were not pleased. They called on security agents to remove the two scientists from the conference venue, their protest banner was confiscated, and their names and abstracts were removed from the official program. A few weeks later, prompted by the AGU committee's response to the protest, the Oak Ridge National Laboratory—where Dr. Abramoff worked—fired her from her research position:[3]

> Right after New Year's, having just returned to work that day, I got the notice of dismissal. I wasn't shocked (they had admonished me for my activism before) but I was definitely sad and surprised. I did not expect they would move so severely and so fast. After I was fired, this made many of my colleagues supportive of our original action, because they saw this was a disproportionate response to what we had done. There were two different letters—with thousands of scientists signing—petitioning AGU to put the abstracts back in the program, which they eventually did.

Scientific conferences are the way academics can share their results and ideas live with other people from their field. Every year or two, hundreds or thousands of scientists might gather in a single location—at times, this might simply be a city's central exhibition hall or a university campus, and at other (less self-aware) times, it can turn out to be a beach resort in the Bahamas or a ski lodge up in the Alps. Once gathered in one place, all these scientists will present on what they've been up to since they last met, through talks and posters describing their work. But most scientists will tell you the fun part of conferences happens outside of these formalities. Lunchtime conversations or late-night bar runs often sprout ideas for new projects and might forge lasting partnerships and collaborations.

Conferences are, therefore, something many academics look forward to. Yet, as of late, they are becoming grim events, particularly when viewed from the outside. To me, this is most salient in the natural sciences. Attend an ecology conference today, and you are guaranteed

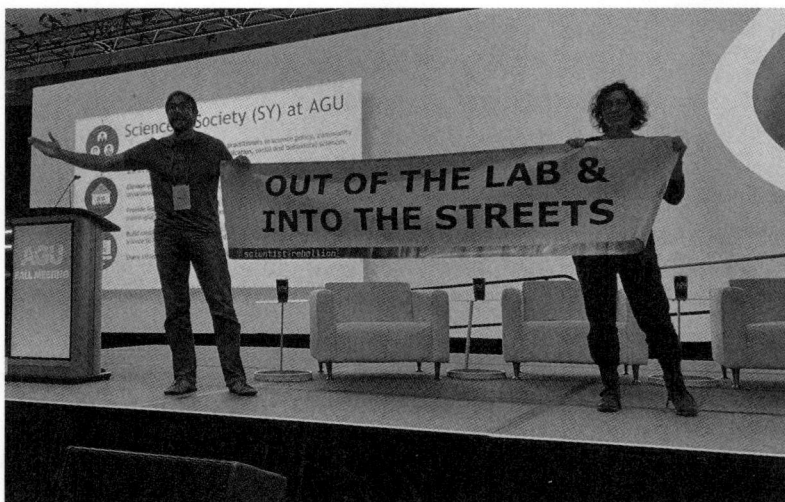

FIGURE 10. Dr. Peter Kalmus (left) and Dr. Rose Abramoff (right) interrupting a plenary event of the American Geophysical Union conference, December 2022. Credit: Dwight Owens.

to see talks about new ways to measure the destruction of ecosystems with ever-higher detail than before, or disappointing results from a last-ditch effort to bring back a species from the brink of extinction. Go instead to a climate conference, and you might witness discussions about which regions of our planet will remain habitable once the ice caps have melted or the Amazon is burnt to a crisp. I have seen scientists attend entire seminars on global biodiversity collapse—and then, once the talks are finished, wrap up the day with margaritas by the beach.

Viewed by someone outside academia, this would seem like madness. To many scientists, it is standard practice. Why aren't there more actions like the one by Dr. Abramoff and Dr. Kalmus? What holds scientists back from acting more, well, logically? In this chapter, I would like to delve into the factors compelling scientists *not* to act, *not* to engage in activism. Why might scientists refrain from stepping out of the lab *in their role as scientists?* What reasons do they themselves give for not doing so? What arguments might other societal actors use to prevent them from stepping out? Finally, and perhaps most importantly, do these arguments make sense?

OUT OF THE LAB: A TWO-STEP PLAN

Some of the above questions can be posed as scientific questions, liable to scientific answers. In an effort to understand the barriers to climate activism, a team of scientist-activists recently published a study in which they used their scientific training to inform and evaluate their mobilization efforts.[4] Lead author Dr. Fabian Dablander told me how they came up with the idea:

> We had countless conversations with scientists about how they see their role on a planet in crisis and what barriers to action they perceive. But a systematic and large-scale investigation was missing. So we put on our scientist hats and got to work.

The authors—many of them members of Scientist Rebellion—wanted to focus on self-perceived barriers to activism: the reasons that scientists give for not participating in actions. They prepared a questionnaire they sent to over 9,000 scientists across 115 countries, asking them if they were willing to participate in activism and, if so, what was holding them back. The results were encouraging: scientists already engaging in civil disobedience made up a substantial fraction (10%) of all respondents, and of the rest that were not, half were willing to participate in some form of activism. Dablander highlighted the main findings from the study:

> It was interesting to see how many people agreed that we need fundamental changes in our societies and felt a responsibility as scientists and academics (not just as individuals) to engage. There's a yearning for doing more.

The survey revealed that self-perceived barriers to activism can be diverse and largely fall into two categories, each requiring different strategies to overcome them:

We distinguished between people willing to engage and people not willing. And then there are people who are willing to engage but not already engaged. In the first step from willing to not-willing, the barriers are mostly intellectual ("I don't know enough," "I don't feel a responsibility," etc.). In those willing to actually engage, it's mostly practical barriers ("I don't have enough skills," "I don't know any activist groups," "I lack the time," etc.). What comes out nicely is this two-step model, and the interventions will be different depending on where a scientist is.

Thus, for scientist-activists, convincing other scientists to act does not have a one-fits-all solution. Intellectual barriers might require intellectual arguments, while practical barriers might be overcome by connecting scientists to activist groups, training them, or helping them build the confidence to engage. In the case of higher-risk forms of activism, like civil disobedience and direct action, practical arguments also revolve around the fear of potential repercussions (of being arrested, of going to jail, of losing one's job, etc.). These could be tackled, for example, by providing information about what might happen during an action: the risks involved, the legal support structures in place, and the various roles (low-risk versus high-risk) that a person could fulfill.

THE SOCIAL PSYCHOLOGY OF PASSIVISM

Beyond self-perceived barriers, researchers have also argued that climate passivism is a collective phenomenon; through a process of "socially organized denial," people refrain from expressing their feelings about climate because they see their peers staying silent as well. In a way, silence spreads like a virus.[5] Psychologist Sally Weintrobe refers to a resulting state of *climate disavowal*, by which we "find ways to remain undisturbed" by the implications of climate change, even after accepting them as true.[6] Rather than frankly confronting the issue, and sharing our doom and dread with those around us, we take distance from the facts and compartmentalize them. It is as if we were all desperately trying to shout "DANGER!" while at the same time unwittingly covering each other's mouths, preventing anyone from saying a word. The silencing, the distancing, and the compartmentalization lead to passivity in scientific circles. If we can't even talk about how climate breakdown makes us feel, then we can't put those feelings into action, let alone encourage others to act as well. Thus, as scientist-activist Dr. Aaron Thierry and colleagues explain, "we are able to intellectually recognize the existence of the crisis without feeling a compulsion to act on it"—effectively a state of "double reality."[7]

Tied to this state is the belief that scientific reporting is somehow enough, that publishing a paper or giving a presentation fulfills the moral responsibility of the researcher toward society and actually contributes to addressing the problem being reported. The belief is that it is enough to just *deliver facts into the void.* In conversation with me, Dr. Julia Basak Halder, a SR member and public health researcher who works with neglected tropical diseases, stressed the psychological toll this belief can take on researchers themselves:

> As I was doing my PhD and the years went by, it became super clear that my research was not going to solve anything and nothing else that anyone was doing in science was solving anything. Nothing is changing. I was getting angrier and angrier. Even though I was comfortable doing science and teaching it, I felt it was inadequate. And so, I started feeling like a fraud. Most scientists got into science to do something good. You constantly hear things like "we're trying to help people through our research" or "our research will contribute to the health of billions of people." But it doesn't seem to add up. I think in my field especially, there is a real sense that we want to do something useful, but there is a reluctance to recognize "maybe we're not doing enough."

Scientific passivism is not just a social phenomenon though. As Dablander and colleagues found, there are strong intellectual barriers against activism as well. Chiefly among them is a belief among many scientists that passivity is a good thing, that it actually makes our science *better.* This belief serves as a logical fortress for the social state of double reality: *It's not just that others around me are not involved in activism; there are sound, logical reasons for them not to do so, and those reasons apply to me as well.*

There are two main intellectual arguments often given to justify scientific passivism. The first one has to do with the notion of *scientific neutrality*, which can manifest both in what scientists do and in how they talk about what they do. The second argument has to do with the risk of losing *credibility* after publicly taking sides on an issue: the fear that we might be less believable as scientists if we also engage in activism. Let's explore each of these arguments in turn.

WHAT IS WRONG WITH SCIENTIFIC NEUTRALITY?

The argument about neutrality goes to the core of what it means to be a scientist. For many, the ideal of a scientist is that of someone who carefully observes or theorizes about the world but remains separate from it. Sociologist Robert Merton, for example, argued that scientists are supposed to act with disinterestedness, that their work must remain independent from their own sociopolitical or personal attributes and desires. Under this paradigm, then, scientists should not let subjective assumptions, emotions, or specific circumstances influence their behavior.[8] Instead, scientists should work toward a universal scientific enterprise, abstracted from all time and place.

Does the Mertonian ideal make sense? Can scientists act with disinterest, removed from their specific circumstances and interests? Should they? If anything, scientist-activists appear to be the opposite of this ideal. It would be hard to describe the actions of Dr. Abramoff and Dr. Kalmus as detached from their specific circumstances and from their interest to protect the Earth system they are studying. Does their activism lessen them as scientists then? This is no small concern for those engaging in action. Indeed, the probability that a scientist will participate in environmental movements strongly depends on how they conceive of the relationship between science and activism, on whether they think activism makes one a better or a worse scientist, and on whether they consider their profession to be consistent with (or even inseparable from) their responsibility to act.[9]

Examining the relation between science and activism is thus crucial for understanding arguments for or against scientist-activism.

For this, we will need to take a small detour and go back to the question posed back in the Introduction. We will need to critically explore how scientists *in general* interact with the world around them, to ask again what scientists really do, how they produce knowledge, and how they share that knowledge with others. I promise that by the end of this detour, we will have a more clear idea of what calls to scientific neutrality really mean and of whether or not they are a sensible argument against scientist-activism.

Generally associated with the Mertonian ideal is the notion that scientists should prevent subjective assumptions from distorting their observations, from influencing their models, and from biasing their conclusions. Scientists should distance themselves from their object of study, lest their proximity pollute the knowledge they are trying to extract from it. This is the first part of the neutrality argument against scientist-activism. In short, it says, a good scientist should—as much as possible—be an *objective* scientist. The second part of the argument goes on to link objectivity with neutrality. In order to be good objective scientists, we must be impartial with regard to the object of our study: we shouldn't attach values or derive emotions from it. Thus, we should not express value judgments or opinions about the world, and we should especially not express them loudly.

Let's break the argument about neutrality down into its components, starting with the first part: the idea that scientists should pursue objectivity—a separation or distance from whatever they happen to be studying. Whether objectivity is actually possible to attain has for centuries been a matter of philosophical debate. Can we ever be truly objective? And even if full objectivity is impossible, should we still strive toward it? Is it a goal worth pursuing?

One of the main criticisms against objectivity (both its plausibility and its desirability) has come from critical and feminist theory. Donna Haraway, for example, sees science as "situated knowledge." She argues that scientists (and any kind of knowledge seeker) can never get rid of the subjective and embodied position from which knowledge is obtained: from the bodies, systems, and machineries that are used to produce knowledge. Claims to the contrary are misleading and ultimately boil down to "a conquering gaze from nowhere"—a way to separate ourselves "from everybody and everywhere in the interests of unfettered power." We should not try to get rid of our position, our situation. Instead, scientists can strive to produce knowledge while being critically aware of the partial perspective through which knowledge is produced. Knowledge seeking is at its best not when we try to erase our circumstances, but when we are conscious of them and take responsibility for them: the particular lenses through which we observe the world, the places from which we do it, and the damage the very act of observation might exert on the world itself. This includes, for example, the "toxic ink and trees processed" that might be used to print our academic papers. Situated knowledge entails moving away from considering the world as populated by "objects" and moving toward recognition of other subjects (human and nonhuman alike) with their own partial perspectives and responsibilities.[10]

A parallel critique has come from decolonial and Indigenous scholars, who argue that objectivity in scientific circles really involves a particular subjectivity, that of the Western white man. Claims to objective science are ways to universalize positions that were originally held by European scholars as their empires subjugated other peoples around the world. In the words of Māori researcher Linda Tuhiwai Smith:

> Research "through imperial eyes" describes an approach which assumes that Western ideas about the most fundamental things are the only ideas possible to hold, certainly the only rational ideas, and the only ideas which can make sense of the world, of reality, of social life and of human beings.[11]

Since the start of the European imperial project and even to this day, other non-Western ideas, systems of classification, and forms of understanding have been erased, undermined, or appropriated as "new discoveries" in the Western canon. Decolonial scholar Boaventura de Sousa Santos calls this a process of "epistemic injustice," which has provided ideological support to (social, economic, racial, and environmental) injustices perpetrated on colonized lands.[12] The elevation of objectivity—the separation of the subject *doing* the research and the object *being researched*—has served to erase or subjugate Indigenous systems of knowledge creation and sharing. It has also disempowered those seen as the research objects. As Tuhiwai Smith puts it, under the Western mode of scientific thought, "the objects of research do not have a voice and do not contribute to research or science."[13]

Thus, critics of science narrowly defined under the Western/Mertonian paradigm argue that one's position—the vantage point from which one looks at the world—can never be universal. What, then, would research that is not made "through imperial eyes" look like? Here, Indigenous ways of knowing can provide an answer. Cree scholar Shawn Wilson explains that, under an Indigenous research paradigm, the nature of reality (in philosophical terms, our "ontology") cannot be separated from how we come to know that reality (our "epistemology").[14] This is because reality is never *out there*, external and separate from the person seeking to learn about it. Instead, reality is best conceived as a set of relationships. To me, a bird song in a forest might just be a pleasant melody or a reproductive call uttered by a member of a species I am studying. To someone living in that same forest, it might be an indication that a body of water is located nearby, or a tip-off that the sun is about to rise. The bird, in turn, might gain knowledge from the human person who is observing. Knowledge through relationships between beings constantly shapes our reality and what we consider to be the "truth." Knowledge is also collectively built, often across generations, never the product of a single individual but always an emergent from connections between beings. And finally, with this notion of relational knowledge comes a responsibility for reciprocity: as Robin Wall Kimmerer teaches in her book *Braiding Sweetgrass*, Indigenous knowledge practices involve a constant giving and receiving—a process of mutual *exchange*, not one-way extraction.[15]

So it seems that objectivity in science is a highly problematic concept: ontologically, epistemologically, and well, ethically. If we try to separate subjects from objects, we keep running into walls. But let's play devil's advocate. Let's say we buy the notion that science *can* be truly objective and that placing ourselves apart from the world we study is ethically justified. Let's assume we can safely do away with Haraway's call for situating our research practices, or the Indigenous critique about the injustice of Western universalism. Even ignoring all this, we still have not faced the second part of the neutrality argument: the idea that neutrality is something a good objective scientist should aspire to.[16] This, I think, is an even harder pill to swallow. Neutrality involves the complete absence of valuation: an emotional and moral indifference to what we are observing, or about the state of the world around us. I have never met a scientist who did not express *some* emotion toward whatever they were studying—be that excitement, admiration, disgust, anger, or even fear. Even those striving for objectivity would be hard-pressed to admit they feel no emotion toward what they conceive of as their study object. Scientists may approach the world around them with scrutiny and skepticism, but they cannot exist outside of it. And so, scientists can never be free of feelings and opinions about that world.

Scientists are also members of a larger social web, and so social values also shape what and how they research, linking the circumstances of the researcher to their work. As Palestinian scholar Edward Said stressed, research is always influenced by the class, beliefs, and interaction networks in which scholars are embedded:

> No one has ever devised a method for detaching the scholar from the circumstances of life, from the fact of his involvement (conscious or unconscious) with a class, a set of beliefs, a social position, or from the mere activity of being a member of society. These continue to bear on what he [sic] does professionally, even though naturally enough his research and its fruits do attempt to reach a level of relative freedom from the inhibitions and the restrictions of brute, everyday life.[17]

I would take Said's argument one step further: our social involvement in society and our beliefs are not just an attachment, an anchor to "brute, everyday life" that we might struggle against but can never get rid of. For many researchers, they are the rudder that steers the boat of their inquiries. Scientists are almost always interested in problems and questions *because* they attach value to them, because they believe they are worth pursuing, because they make them feel something.[18] Shawn Wilson quotes another Indigenous scholar—Eber Hampton—who says that "we do what we do for reasons, emotional reasons. That is the engine that drives us."[19]

Some scientific fields are impossible to dissociate from the values and belief systems that engendered them. They can never be neutral. The medical sciences are a prime example: a giant body of research developed in order to preserve human life, mitigate suffering, and evade disease. Cancer researchers are rarely interested in cancer for its own sake—they want to help cure cancer (or at least the type of cancer they may be studying), not help it spread. Many so-called crisis disciplines—like epidemiology, public health, and conservation biology—explicitly take a stand for the conservation of life in the face of destruction.[20] As Julia Halder put it, in these fields, there is a sense that scientists want to do *something good* or *something useful*, that their work is supposed to positively contribute to society. Barring Hollywood movie villains, I would be hard-pressed to find a conservation biologist set on exterminating life on Earth as fast as possible.

SCIENTIFIC NEUTRALITY IN PRACTICE

The argument about neutrality acquires even more holes when we look at how science is actually practiced. Funding agencies, university governance bodies, state governments, and societal values constantly influence the work researchers perform inside their laboratories. Some of the worst atrocities in history were carried out with the approval, collaboration, or direct involvement of scientists and scientific institutions, whose actions were shaped by the politics and values of their time. Even with all its marvelous achievements, science (as Paul Feyerabend reminds us) "often produces monsters."[21] The 1940s saw myriad forms of experimental torture committed by German medical researchers inside concentration camps.[22] Concurrently, but on the other side of the world, the infamous "Unit 731" program—

carried out by Imperial Japanese Army researchers—is estimated to have killed over 200,000 people through biological weapons testing and vivisection.[23] And just a few years after the Second World War, the United States funded and directed an experiment in which doctors inoculated around 1,300 Guatemalan residents with deadly diseases. The unwilling test subjects included orphans, mental health patients, and prisoners who were injected with syphilis, gonorrhea, and chancroid without their informed consent, all in the name of science.[24] Thus, human torture and experimentation were very much within the scientific playing field in the 1940s and 50s. Scientists were not just allowed to commit these atrocities, but paid by their governments for carrying them out. Yet, in the present day, it would be hard for me to find a funding agency supporting research into more painful ways to cause torment or unwillingly experiment on human beings. This is not because such research would not be novel, impactful, or even insightful, but because prevailing societal norms (thankfully) treat such activities as highly unethical and not worth striving for.

Neutrality is thus not a defining feature of how scientists operate in the world. More than that, the imperative for scientists to be neutral assumes there always exists a *neutral position* that is absolute and self-evident. In reality, the neutral position effectively boils down to the current status quo—the way things are today—be that in issues of climate, social, or economic justice. Telling scientists to be neutral is just a roundabout way to tell them to stay in their lane and not question society as it is currently structured. As Dr. Abramoff explained:

> We've gotten too good at following the rules. We are no longer willing to follow our hearts as much. We're trained to be conservative, unemotional, unbiased—when you apply these values religiously, it basically means we're acquiescing to the status quo. And the status quo right now is a deadly state of being. It is interesting to me that some of the great discoveries in history have almost all been accidents. Somebody tinkering with something that they shouldn't have been tinkering and noticing something interesting. Without us going rogue and giving ourselves the permission to be more experimental, we're not going to learn anything new. And if we don't tinker with our social systems, we're not going to learn anything new about how we can save our society.

Direct human torture is now generally outside the political limits of university research. Yet, other types of work causing suffering are still permitted. Experimentation on animals is one example; another is fossil fuel research. As we will see in later chapters, projects to develop more extractive ways to obtain oil, coal, and gas are still within the legally sanctioned roster of scientific activities. At the same time, they are also increasingly called into question: demands for excluding activities of this type from campuses worldwide (for putting them in the same bag as research on human torture) are steadily growing.[25]

THE EVIDENCE SUGGESTS THAT OUR HOUSE MAY BE ON FIRE

Beyond the practice of research, arguments in favor of passivism are often made in the context of scientific communication. Researchers are instructed to be detached in the language they use to disseminate information about their work, to avoid emotional wording when reporting on the results they obtain and their implications. They should be mere conveyors of abstracted knowledge: it is the rest of society that is responsible for deriving values from that knowledge.[26]

This argument is also based on neutrality: it assumes there is a "neutral" scientific language, and researchers should stick to it. Yet, emotionally detached research manuscripts are just as value laden as emotionally charged ones. The latter convey that the topic at hand is worthy of immediate and urgent attention, while the former do not. Emotionally charged writing is thus much more likely to activate others to take action in times of crisis.[27] Emotion also shows that a scientist has wholly interpreted and comprehended their results; it reflects a deeper understanding about research, in the full context of its societal consequences.

Dr. Abramoff and Dr. Kalmus are choosing to follow unconventional, emotionally charged means of communication to convey the implications of climate collapse. So are the scientist-activists blocking roads or occupying buildings while calling for a "climate revolution."

But this makes them better scientists, not worse. The emergency we are living through is no longer something that can just be conveyed in a scientific paper. When the house is on fire, we don't send a letter to those inside; we desperately wave our hands and shout "WATCH OUT!"

In the film *Don't Look Up*, a team of astronomers discover a giant meteor is heading toward Earth, with the potential to destroy all life on the planet. During the movie, it is the scientists—those equipped with the greatest amount of information about the meteor—who push the most for political action. They begin by politely asking politicians in power to do something, and they realize too late they should have been riling up the people instead, with much more urgency and desperation than they had first foreseen.[28] Ultimately, their efforts fail, and they must face the consequences of this failure: the end of human life on Earth, including their families' and their own lives, as the meteor's trajectory (unsurprisingly) follows their original predictions and crashes into our planet. As director Adam McKay tells it, the movie is a metaphor for inaction in response to the climate emergency and for the ongoing banalization of decision-making.[29] But it goes deeper than that: it is also an illustration of what can happen when scientists act so "neutrally" they end up shooting themselves in the foot.

A RISK TO CREDIBILITY?

Beyond neutrality, the second main argument thrown against scientists doing activism has to do with their credibility. Even if it is impossible for them to be neutral on an issue, critics argue, they shouldn't adopt open stances that favor an opinion or policy. Otherwise, they risk compromising their credibility with the public.[30] This is a more practical argument than the one about neutrality. It argues that taking a very public stance might limit the effectiveness with which scientists can communicate with the rest of society. A synthesis of a colloquium on scientific engagement summarizes this view: "[If] the science community is perceived as having a 'values-based' agenda, scientists put themselves at risk of losing their credibility and, thus, lessen their potential impact on policy."[31] Thus, in order to affect policy (presumably in some direction?), scientists should look like they are free from values. Effectively, to be non-neutral, we must, for all intents and purposes, look like we are neutral.

From personal experience, I can say the fear of credibility loss is a fear many in the academy hold in good faith. Faced with this argument, some scientists have conducted studies to assess whether it actually holds water. For example, Kotcher and colleagues conducted a randomized survey of US residents, across various demographic categories including age, sex, education, region, income, and race. The survey participants were then asked to score the perceived credibility of a fictional scientist making various public statements in favor of environmental policies. The authors found that the credibility of the scientist was high in response to all but one of the statements made, and the level of trust in climate science remained high across all survey statements.[32] A similar study—conducted on a sample of 2,453 US-based respondents—showed that scientific credibility is not necessarily hurt by scientists' adopting particular positions on an issue, and it can—under certain circumstances—actually be increased by taking sides.[33] Yet another study—conducted via a survey of German and US citizens—points to a similar conclusion: though openly standing in support of particular policies might negatively affect the perception of scientists as "objective," this behavior does not decrease the trust others place in them. The researchers also found that the public generally believes scientists should engage in climate-related issues and that their efforts at doing so are currently not as large as they should be.[34]

The results from all these studies suggest the credibility argument is not really supported by the evidence: scientists are not unlikely to lose trustworthiness when taking positions in societal debates. Indeed, they are often expected to do so by the general public. This has an important corollary: if the public expects scientists to engage, then adopting behaviors that avoid or sidestep such engagement might actually backfire.

What happens when scientists are too passive in a crisis? On April 6, 2009, a 6.3-magnitude earthquake hit the Italian town of L'Aquila. The earthquake caused an estimated 309 deaths, and approximately 1,600 people were injured. The quake's intensity was not particularly extreme, but its reach—in terms of human life—was highly unusual. So what happened? A follow-up trial helped uncover errors in scientific communication before the event. Tremors had been felt in the days leading up to the earthquake, yet experts repeatedly reassured the population a deadly event was not likely. This caused people to not take precautions, like preemptively leaving their homes.[35] Admittedly, earthquake prediction is plagued with uncertainties—as it was later argued by the seismic scientific community, who were outraged by the accusation that scientists had underplayed the risks. Yet the scientists knew a risk was present, and they chose to minimize it, rather than amplify it, in order not to alarm the population. Sometimes, being overly cautious is the most dangerous path a scientist can take.

Fears of seeming "too alarmist" have led researchers to underreport the impacts of climate collapse.[36] In a very real sense, this has led to preventable loss of life. The public continues to be unaware of how terrified scientists—those closest to the data—really are. Behind closed doors, many scientists already assume much more catastrophic scenarios than what one would expect from a reading of the dryly written IPCC reports. In 2021, an anonymized poll showed that the same authors of these reports are skeptical that governments will follow their recommendations, and many privately believe the worst scenarios will become a reality within their lifetimes.[37]

One final argument related to credibility revolves not around whether scientist activism is worthwhile, but around *who* is allowed to do it: *you're not climate scientists, so you should not be out on the streets calling for climate action.* This argument states that one must have an extremely specific expertise to credibly engage in protest. This expertise is usually defined narrowly enough as to be meaningless and to render collective action impossible. Should only climate scientists studying ocean circulation be allowed to protest sea level rises? Should tree botanists be the only ones allowed to occupy the

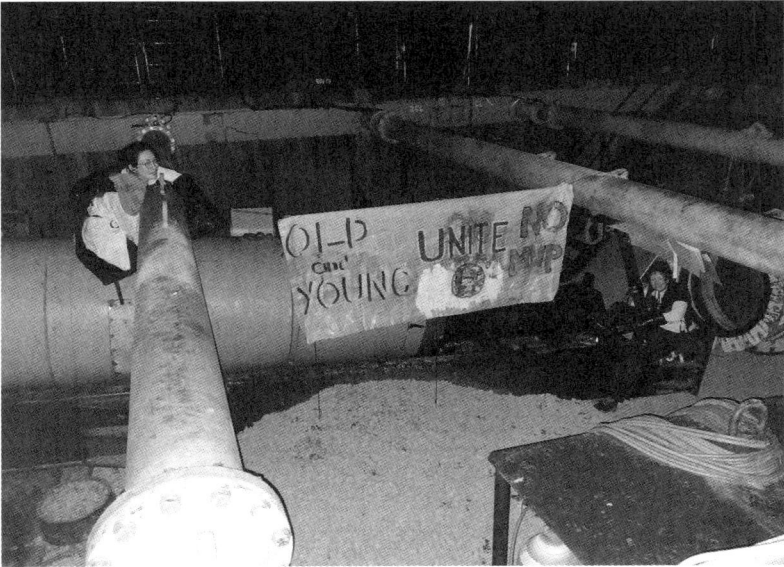

FIGURE 11. Dr. Rose Abramoff (left) and a member of elder climate group Rocking Chair Rebellion, Marty Zinn (right), chained to the drill equipment of the Mountain Valley Pipeline, September 7, 2023. Credit: anonymous / SR Turtle Island.

offices of a logging corporation? Accepting this argument robs all scientists of agency: it is siloing taken to the extreme. Those who espouse it would want us to stand alone, each on our small little island of individualized resistance, separate and distant from all our colleagues.

To counter this last argument, scientist-activists insist that knowledge about climate breakdown is brought about through a vast network of cooperation, that climate is but one component of a vast interrelated social, economic, cultural, and ecological system to which all scientific (and nonscientific) backgrounds can contribute. They also argue that it is not necessary to be a climate expert to understand the basic tenets of climate breakdown. Ultimately, the demand that only "real" climate scientists do climate activism is an ideological construct, created to delegitimize and demoralize scientist-activism as such. Ironically, I have also heard the reverse argument being used by some scientists to justify their own passivity: *yes, scientists should be out on the streets demanding climate action, but I am a climate scientist, and so I would risk losing credibility if I were to do so myself.* Siloing thus works not just as a deterrent, but also as a defense mechanism, an excuse for staying passive.

THE RISE TO THE TOP

In summary, scientist-activists like Dr. Kalmus and Dr. Abramoff are not just breaking with the convention of the "double reality": they are following the vast amount of evidence indicating that credibility is not lost when one steps out of the lab. And they are actually gaining credibility through their actions. Since joining SR, Dr. Abramoff has been arrested over seven times and has gathered thousands of followers on social media. Some of her actions are particularly daring: she has chained herself to the White House fence to demand President Biden declare a climate emergency, and once, she locked on to the drilling equipment of the Mountain Valley Pipeline in West Virginia, together with members of the elder climate group Rocking Chair Rebellion. Abramoff thus became the first scientist in the United States facing felony charges for climate resistance.[38] Meanwhile, Dr. Kalmus made global headlines for chaining himself to the Los Angeles headquarters of Chase Bank, one of the biggest financiers of fossil fuels in the world. Kalmus did so while facing dozens of police officers in anti-riot gear, who surrounded the scene.[39] He is now the most followed climate scientist on Twitter/X. Thus, these scientists are embodying critical perspectives on scientific outreach— breaking with the notion that scientists should be neutral, emotionless, or dislodged from the world that they study.[40]

Yet, for them as for many other scientist-activists, convincing other scientists to join the movement remains a challenge. Should we use rational arguments to explain why activism does not conflict with our role as scientists? Should we appeal to evidence pointing against damage to credibility? Or should we try to tap into people's emotions? Dr. Charlie Gardner explained to me that the answer might not be as simple as one or the other, that the different barriers to activism are interrelated:

> The rationales we tend to hear from scientists themselves are not necessarily the true reasons why they are not joining movements. It sounds a bit selfish, and self-centered and perhaps self-promoting, to say "I don't want to get involved in activism because I need to focus on high-impact papers," so you can't say that and you can't admit it to yourself, and you come up with other reasons to allow you to excuse yourself. It's no surprise we want to stay on our chosen paths. Many of us have had this ambition since we were young. It can be almost our sense of identity. "I'm a scientist. I've always imagined a future of being a respected professor." And in academia, the people that rise to the top usually share one characteristic: they are able to say no to distractions and focus on the things that are going to get them to the top. So of course your brain is going to play tricks on you to defend that dream, and it does so in all sorts of ways.

Perhaps, then, it is worth taking a closer look at the structures that incentivize the "rise to the top" and the competition for "high-impact papers" in scientific workplaces. In the following chapter, we'll explore the ways through which structural academic logics incentivize passivity, and the approaches scientist-activists are adopting to respond to them.

CHAPTER 5

Academia and its Discontents

" . . . the great adventure of human curiosity presented to
them as children is replaced by the theme of a vocation that
demands body-and-soul commitment. And this is what we
accuse today's young people of no longer accepting: compli-
ance with the sacrifices that service to science demands."

—Isabelle Stengers[1]

Lou studies glaciers for a living. More specifically, she studies the
invisible organisms that live on top of them. Microscopically tiny
algae and bacteria manage to survive, even thrive, on the surface
of the ice. For her, these organisms are particularly fascinating
because they can "color" the natural whiteness of glaciers, decreasing their capacity to reflect light from the sun. She is pursuing a
PhD to understand how these little creatures can contribute to
glacial melting, which is a crucial component of our climate
system.

Her research takes her to some of the most majestic landscapes in
the planet—realms of pristine snow and frozen crystals far up north
in Greenland and Norway. She collects her measurements while
camping on the ice in the summer season. At her campsites, she
wakes up and goes to sleep under the constant presence of the sun,
which never sets at that time of the year.

But these landscapes are dying fast. Many of the glaciers Lou has
visited will disappear within her lifetime, she told me:

It's difficult to just look at a glacier and know that it will not be there within a few decades, knowing that so many creatures that depend on the ice will also disappear. It fills me with sadness but also anger, because of the injustice of it all: the natural world is going through changes that it didn't ask for, affecting our climate and all existing beings. The disappearance of snow and ice means the extinction of species as well as the extinction of very ancient cultural heritages in the Arctic.

Today, Lou is not on the ice though. She is instead harnessing her anger into something else: taking a break from her research on glaciers to go face the cause of their impending death. She has traveled south, to central Germany.

It is January 14, 2023. The tiny German village of Lützerath—nicknamed "Lützi"—is today one of the main sites of resistance against the advance of the fossil fuel machine in Europe. Lützi sits right on the border of an open lignite coal mine, hundreds of hectares in size. The multinational energy company RWE means to destroy the village so they can expand the mine even further.[2]

For months now, Lützi has been occupied by hundreds of activists, locking on to buildings and trees to resist the expansion of the megamine. Their motto is "Lützi bleibt" (Lützi stays). They don't just want to preserve the village; they want RWE to keep the coal in the ground. Lignite coal is the most polluting of all fossil fuels, due to its low heat content. At just 17 megajoules per kilogram, it is half as efficient as coal from bitumen, so one needs to burn a lot more of it to obtain the same amount of energy.

Over the past week, RWE and police in riot gear have moved to forcefully evict the activists who still refuse to leave. Giant bulldozers have entered the village and are working day and night to destroy the town buildings.[3] As a Scientist Rebellion member, Lou is joining about 35,000 people who are now surrounding the town, in a last stand to retake it. She watches in disbelief the desolation all around her: kilometers and kilometers of barren soil, which end just at the town's limits. Once dense forests and farmland, the entire landscape has turned into a monochrome beige basin, all dug up using giant bucket-wheel excavators, over 14,000 tons of metal—the

FIGURE 12. The open-pit lignite coal mine Garzweiler II, during the eviction of Lützerath in January 2023. One of the bucket-wheel excavators is in the background, center-right. Credit: Lou.

largest machines in existence—designed to destroy anything in their path.

> It's heavy, seeing these huge machines digging and digging, and taking so much from the land. The sound, the visuals. The evil is really there, in front of your eyes.

Though Lou and the other activists try to reenter the village, they ultimately fail to save it. Hundreds of heavily armored police agents encircle Lützi, enabling RWE's machines to continue their work.[4] The bulldozers clear the last vestiges of the buildings, and the excavators move to dig the ground underneath. Writer Peter Gerderloos compares this type of mine landscape to Tolkien's Mordor—a vast wasteland as far as the eye can see, where nothing is alive.[5] As Lützi vanishes into rubble, the wasteland looms ever larger, the machines encroaching on everything and everyone along their path.

In her dual role as nature observer and nature defender, Lou is a rare case in academia. She has seen firsthand both the causes and the consequences of climate change. But she doesn't feel it is enough to just see. She also feels compelled to act. Her analysis on glacier decline is complemented by her fight against the companies that are causing it. A few days after her stay in Lützi, Lou comes back to her lab, but something deep inside her has changed. It feels wrong to just continue doing science as usual, knowing the fossil machine is still out there, still advancing. She feels a stark contradiction between her two identities—the scientist and the activist:

> I increasingly think that academia (in its current form) is part of the problem. Even if we don't produce money or goods en masse, we're still part of the same machine, consuming tons of resources to produce manuscripts, papers, and reports, without really thinking whether the knowledge we're creating is even being used properly. It all serves to keep the gears of capitalism moving—the same gears I was trying to escape from, when I first decided to become a scientist.

Sitting with her for a chat, I ask Lou if she talks about her activism to other members of her lab. After all, one would think that a research group specializing in climate change and glacier collapse would be buzzing with conversations about Lou's experiences in Lützi. That couldn't be farther from the truth:

> Even though we work directly on the effects of climate change, there's no discussion about climate activism or climate politics in our research group. It's very difficult to even start a conversation about these topics. You can't really talk about it, there's no culture of discussion about this in my field, and it's especially difficult if you're an early-career scientist.

What are the reasons that make it so difficult for Lou, as well as other scientists, to open up about this activism? How did academia become an environment in which even *talking* about activism is frowned upon? And how are scientist-activists navigating and challenging this environment?

THE RULES OF ACADEMIA

Academia is a system, and like any system, it has rules of operation. These rules determine how the system works: what it takes in, how it modifies and transforms it, and what it produces as a result. These rules are not immutable: they are subject to constant negotiation, and they can change and evolve over time as the system evolves.[6] Some rules are explicitly stated—for example, in academic handbooks, institutional documents, speeches, deans' orations, or on university websites. Other rules are tacit—hidden from view, but known to all academics. The latter are perhaps the most prevalent, so ingrained in the system they are seemingly unmentionable, let alone questioned.

So what rules apply to academia? And how do they influence how scientists behave within this system? We scientists are curious creatures: we seek to understand how the universe works. But, like any other creature, we need to eat, drink, sleep, and have a roof over our heads. As we try to do our research, we also seek the means of our subsistence, which, under capitalism, require a salary. The essential rule that determines how this happens in academia can be summarized like this:

Thou shalt obtain grants, or else perish.

Grants are short-term disbursements of money that pay for research equipment, publication costs, a lab's running expenses, and—crucially—salaries. Grants can come from the state, NGOs, foundations, or corporations, and they are obtained through cutthroat competition between researchers. More often than not, past track records play a large role in who gets them. If you did well before, you will do well in the future. As a general rule in academia, the rich tend to get richer.

Scientific track records are measured via "metrics of evaluation": scores that quantify how productive a scientist has been so far, how many papers they have published, and whether those papers have made it into prestigious journals. The threshold for what counts as prestigious comes from other scores, designed by the corporations that produce these journals, in an effort to push academics to publish in them. Scientists have taken the bait: over the past decades, they have come to believe that some journals are more prestigious than others, and the collective belief itself creates the prestige.[7] Today, scientists are evaluated in hiring and grant committees based on these badges of artificial worth.[8] Frequent publication in "high-profile" journals thus improves one's prospects of landing a permanent faculty position. These positions are highly coveted, as they have been made scarce by a relentless process of precarization.[9]

The pressure to publish in commercial journals has allowed publishing corporations to take advantage of their own content creators. Most academics usually pay to publish their own manuscripts. The money for this may come from the research grants scientists might hold, university funds, or (if they don't have either) scientists' own pockets. These are not meager amounts. Because certain journal names have become marks of prestige, corporations have been able to ramp up prices to extremely high levels. If you want your research manuscript to appear in high-status venues like *Nature, Cell*, or *The Lancet*, you will need to pay thousands of dollars. This is well beyond the actual operating cost of labor and resources that an article needs in order to be published,[10] so scientific publishing is a highly lucrative business. In the past two decades, major commercial publishers have reported profit margins hovering in the realm of 25 to 40 percent.[11] Unsurprisingly, the magazine *New Scientist* has recently called academic publishing "the most profitable business in the world".[12]

Scientists who can afford it remain willing to pay because it means a chance of landing a better job, or even having a job at all. This has led to hyperinflation: the rise in publication prices is now much higher than the average inflation rate in Europe and North America.[13] In fact, the price to publish in the most prestigious journals exceeds the monthly salary of almost any academic.[14] When not charging researchers, corporations charge readers by putting articles behind paywalls. This, in turn, is siphoning resources away from public universities and libraries that pay for these subscriptions in bulk so their members can access the content inside the closed-off journals.[15] Academics are really the middlemen in these transactions: billions of dollars are being siphoned away from the public—through academia—into the pockets of shareholders.[16] The rich get richer again.

Even with this knowledge, our mentors and supervisors often advise us to publish frequently and prestigiously—or otherwise risk damaging our future job prospects. This is the second law of academia:

Thou shalt publish, or else perish.

This mandate looms large in our minds. The pressure to publish generates an environment of hypercompetition, in which insecurity is high and trust is low: the perfect recipe to discourage critical thinking and collaboration[17]—but a sure way to prevent dissent.

How did this come about? Education scholars Sheila Slaughter, Larry Leslie, and Gary Rhodes have argued for a theory of "academic capitalism" to explain how campus research, teaching, and administration have been transformed through the neoliberalization policies of the 1980s and 90s. They see the hypercompetitiveness that scientists are subject to as part of a broader project that, at the turn of the century, plugged academia into global capitalist circuits. The project has blurred the lines between the scholar and the market, conceptualized scientists as "entrepreneurs," and bound knowledge sharing to market profitability:

> The academic capitalism knowledge regime values knowledge privatization and profit taking in which institutions, inventor faculty, and corporations have claims that come before those of the public. Public interest in science goods are subsumed in the increased growth expected from a strong knowledge economy.... Knowledge is construed as a private good, valued for creating streams of high-technology products that generate profit as they flow through global markets.... These models see little separation between science and commercial activity. Discovery is valued because it leads to high-technology products for a knowledge economy.[18]

In a similar vein, sociologists Kleinman and Vallas argue that the practices and norms of behavior of capitalist industry have been rap-

idly seeping into academia.[19] It's not just that science is now supposed to work in the service of corporations; it is supposed to work
like corporations. Metric evaluation is a prime example. It was
adopted from corporate productivity evaluation practices, which
tend to incentivize competition: collaborative work is undervalued,
and scientists are expected to climb the academic ladder at all costs.[20]
At a higher level of the hierarchy, managers compete in their own
metrics games against each other: rectors, presidents, provosts, and
deans are constantly seeking to place high in rankings of academic
"impact" and "efficiency," as if they were CEOs fighting for a bigger
share of the market.[21] This is not just a simile; as universities increasingly rely on private foundations and student fees for financial support, market competition becomes the true goal. University rectors
and presidents are not just *like* CEOs; for most intents and purposes,
they *are* CEOs.

Ultimately, under academic capitalism, those producing the largest output at the smallest cost make it to the top—*output* here meaning the number of papers in prestigious journals, or the number of
patents, or the number of scientists who are good at producing such
papers and patents at a high rate. The system is geared toward productivity and competition—instead of intellectual freedom and crea-
tivity.[22] And so, rather than fostering critical analysis, the system
rewards scientists for publishing more, faster, and better. Peter Higgs,
Nobel Prize winner and discoverer of the Higgs boson back in 1964,
famously said that the pressure to "keep churning out papers" would
have meant that no university would have employed him in today's
academic system.[23]

What's worse, given the high degree of competition and mistrust,
most scientists keep their complaints to themselves. With so few permanent positions in the job market, and so many wanting to get one,
scientists play it safe and follow the metrics game. If you don't want
to play along, someone else will replace you. As philosopher Isabelle
Stengers has pointed out, the curiosity, the drive to question that
once drove young people into science, is thereby suppressed. They
are eventually "reduced to the sadness we call depression, or end up

as the kind of opportunistic cynics who know how to make all the right moves."[24] This brings us to the last important rule of the academic system:

Thou shalt not be seen as too political.

The engagement of scientists in the questions and problems of today's society—be that via advocacy or activism—contributes nothing to performance metrics, or to the university's rankings. These activities are seen as a distraction at best, and a dangerous risk at worst.[25] So, universities generally discourage or invisibilize them,[26] and they try to keep academics busy on things that can be measured, quantified, and commercialized.[27] Thus, science and scholarship are predominantly practiced in a system that not only stifles cooperation, but also deters engagement with the wider world. The silence in Lou's lab is not the exception, but the norm: staying quiet becomes a strategy for success and, eventually, a tacit rule everyone is expected to follow.

THE POSTDOC VS. THE DEAN

Some scientist-activists are now challenging this system—even on its own terms. This is ultimately part of the ethos of Scientist Rebellion: to rebel against the norms of our institutions. We already saw the example of Dr. Rose Abramoff, risking her own job to urge academics to take to the streets. Another system challenger is Dr. Sanja Hakala—an insect biologist at the University of Fribourg. In her efforts to mobilize for scientist-activism, she has stood up against those at the highest rungs of the academic ladder in her place of work.

Sanja studies social insects: bees, ants, and termites. Her research dwells on how these little bugs organize together to build giant complex structures, like the intricate tunnel systems of leaf-cutter ants. She is a postdoctoral researcher, a "postdoc": one step above Lou on the academic ladder, but still a temporary position with a fixed term length.

By the second half of 2022, Sanja had become heavily involved in climate activism. She had joined various Swiss climate movements, including Scientist Rebellion and Renovate Switzerland. The latter was at the time making headlines for their nationwide highway blockades. They were putting pressure on the Swiss Federal Council to implement a national plan to insulate 1 million homes by 2040—a way to both mitigate carbon emissions and lower energy costs.

It was a strain on Sanja's already busy time: on top of writing grant applications to continue her research (and remain employed), she was also connecting activists with academics, advertising for mobilization events, and even participating as a blockader in the highway protests. Like the social insects she was studying, Sanja was trying to get people organized, to make them realize they could achieve a lot if they just coordinated efforts. And she was doing all of this in her free time.

Sanja did not see her activism as separate from her scientific outreach. The groups she was working with were trying to amplify the science of climate change, after all. They were trying to do—via unconventional means—what she thought all scientists should be doing: sounding the alarm as loud as they could. And yet activism was not something that was recognized by her university as a valid form of work. She wanted to poke the university about this, to challenge its commitment to passivity. She decided to organize a panel symposium in her department, open to anybody in the community who would be willing to attend.[28] Sanja invited an IPCC author, a philosopher, a finance professor, and an activist from Renovate Switzerland. The four speakers were to gather together with other academics to discuss a simple question: *Should scientists be doing activism?*

Swiss academia is very hierarchical, with professors having the bulk of the power. Not only are professors the only ones with permanent employment, but they also have large amounts of power over what those below them can do. As a postdoc, Sanja found she was even restricted from booking a lecture hall by herself so that the symposium could take place. Nevertheless, she persisted and finally managed to get the one room on her campus that anybody could book.

When she started advertising the event across campus, she was met with yet another obstacle—a powerful professor disliked the topic of the debate and tried to get it canceled, she told me:

One professor in my department saw one of the posters for the event and completely flipped! In the poster, there was a photo of a climate activist sitting on a road (she was one of the panel participants), and that apparently set him off. He discussed with other people how this postdoc who is not even supposed to be doing anything on her own is encouraging illegal activities. He tried to get the symposium canceled by emailing the dean and CCing my boss behind my back.

The dean then sent Sanja a list of things that she and the other participating academics were not allowed to do in the symposium, including calling others to join in civil disobedience, or even asking scientists to do activism. He told her she must not use the symposium as a platform for recruiting people for activist groups and that she could only use the email addresses she collected for invitations to other academic events. This way, the dean was explicitly drawing a line between what can be considered "academic" and what can be considered "activist." By the dean's definition, academics were meant to be passive actors, refraining from participating in the civic life of their society. Activist events could not be "academic," even if those participating in those events were—indeed—academics.

Sanja was not disheartened. Indeed, it was the dean's definition of what can be allowed to happen in academia that the event was meant to question. And what could be more academic than exploring the very concept of what academics are supposed to do? She also knew it was within her right to organize a symposium to discuss topics of relevance to society. That is ultimately what academic freedom boils down to: to be able to speak truth to power, in the search for greater understanding.

Sanja doubled down on the event, using it as a way to challenge the university's response:

> We did the event anyway and talked about the censorship we were receiving from the university leadership. The audience was shocked about the dean's behavior. Other professors and university employees then stepped up in favor of free speech, sending letters to the dean in support of the event.

This, in turn, triggered a wave of discussions about activism in the hallways and classrooms of the University of Fribourg:

> News about what had happened grew across campus. The university
> communications department was on our side. They ended up publishing
> lots of articles about activism and academic freedom and even organizing
> an outreach event on this topic for the general public. We got a lot more
> visibility, just because people were trying to prevent it from happening!

The censorship Sanja received ended up amplifying her message. She was even interviewed by the university newspaper,[29] which ran a series of articles on free speech and civil disobedience, prompted by the symposium.[30] At the end of the day, she got to tell the entire campus what she thought, why academics had so far failed in their communication strategies, and how they could embrace civil disobedience as an effective way to connect with the public. She ended her interview with a resounding statement: "Let's face it: it's not enough to just pretend to create impact."

THE ONES WHO WALK AWAY

For some academics (especially those with permanent contracts), it is actually possible to refocus one's daily activities so as to support climate activism. Professor Adam Aron at the University of California, San Diego, is one such case. After 20 years of working in neuroscience, he decided to completely abandon his original career focus. He instead founded the Climate Psychology and Action Lab—a research group dedicated entirely to understanding the main motivators that drive people into climate activism, or prevent this from happening.[31] Through his research, Prof. Aron is now trying to help environmental groups increase in numbers and become stronger.

For others, particularly young researchers with temporary contracts, academia provides more roadblocks than opportunities. Rather than trying to reform the system from within, some scholars choose to step out of academia altogether and embrace activism full-time. This is no easy decision: without an actual salary, full-time activists rely on "voluntary living expenses"—small, short-term sums of money that come from donations made to climate groups.

Activists without a stable influx of money may find themselves in more precarious living conditions than PhD students or postdocs.

The poster photo of the activist that got Sanja into so much trouble was actually a photo of Dr. Anaïs Tilquin—a scientist turned full-time climate activist. Anaïs originally performed research in global ecology and ecosystem restoration. As a postdoc, however, she found it more and more difficult to concentrate on her work. Through the climate activism she was doing on the sidelines, she was seeing firsthand how politicians were ignoring the knowledge she and others like her were generating. She expanded on this in an interview I had with her:

> I decided that the climate emergency was something I wanted to dedicate myself to full-time. And yes, I believe in fundamental science. But right now, some things are more urgent and useful than others. We're rapidly losing everything to climate breakdown. My research didn't seem useful anymore. We can't restore something that no longer exists. And sure, science is good because it allows us to accumulate more knowledge. But the bottleneck to fix today's major problems is not knowledge, but political opposition.

As a full-time activist, Anaïs has helped organize major city occupations with Extinction Rebellion, as well as highway blockades with Renovate Switzerland. She has also served as spokesperson for these movements; videos of her often appear in Swiss media.[32] She is unassuming and determined in her demeanor, and she speaks with clarity and frankness. Her decision to fully transition into activism did not come without costs:

> I've sacrificed a lot of personal projects and relationships to my activism. But I don't really see them as sacrifices, because I don't see I have a choice. Unless I talk to people with similar scientific profiles as me: those who know how bad things are, who know that social movements are the key to our predicament, and who still remain inactive. It feels like a sacrifice because some people are not making it, and it hurts. I also feel rage and disgust. I was expecting so much better from my community.

Feelings of disappointment with fellow colleagues run through many of the scientist-activists I have interviewed, including Sanja and Lou. They all expected more from their mentors and coworkers, especially those who hold powerful positions in academia. Anaïs blames the systemic incentives to be passive, but she also highlights the privilege of those who can afford a career in science:

> The way our scientific system is organized is uncooperative—and as competitive as can be. People therefore end up being very paranoid about their careers. Young scientists claim they don't want to take risks so as not to jeopardize the next step in their careers. And older scientists claim they are too busy, that they don't have time for activism: "Yes, if I were younger, I would do it." But the problem is also a lack of diversity among scientists. Those doing science here in the Global North are mostly people from privileged backgrounds, people who don't realize how easily things can go sideways. We've been sheltered and we don't understand how we got the societal privileges we have today.

Anaïs gave me an illustrative example to make her case. In November 2021, she was providing support to Guillermo Fernandez—a father of two who had decided to go on a hunger strike in front of the Swiss Federal Assembly. Guillermo's was a hunger strike without a term limit. He was prepared to continue fasting unless the assembly members received a day of training by researchers from the IPCC and the IPBES (a platform similar to the IPCC but focused on biodiversity loss). His reasoning was that, once trained, the politicians would no longer have an excuse of "not knowing" enough about the climate and ecological emergency and would therefore take much-needed measures.

Throughout the strike, Anaïs was making sure Guillermo's plight would not go unnoticed. As she explained to me, the logic behind a hunger strike is to light a fire under the feet of your allies. A hunger striker's opponents might remain indifferent, but their supporters now have a strong motivation to move decisively and actually do something, to make sure the demands are fulfilled before the striker dies. Anaïs was one of these allies, and she was trying to get other scientists to be on board as well:

> During the strike, I remember being in a Zoom call with IPCC and IPBES authors in Switzerland. I was trying to get them to sign a letter to the assembly saying they supported Guillermo's demand, that his claims were consistent with the latest recommendations from these international bodies, and that they were prepared to give the proposed climate training as

soon as they were authorized. In the end, we got them to sign the letter, but several people still said things like "I'm so sad for the Global South, but well, us here in Switzerland are going to be fine." One IPCC scientist towards the end of the call even said, "I'm having second thoughts about this letter: these demands are so obvious we should have made them earlier and we didn't—we would look like unorganized scientists if we do it now." He was questioning sending the demand letter because he and the others might look bad. He was considering letting this person die to save face!

The assembly eventually caved, and Guillermo ended his strike after 39 days of not eating any food. The training finally happened the following May. Since then, Anaïs has helped Guillermo find other ways to do activism, but he still fears for his children's future. "He's one of my favorite people in the world," says Anaïs. The experience with Guillermo strengthened her commitment to climate activism. It also showed her how even the most renowned of all climate scientists may still be clouded in their judgment by the pursuit of prestige.

THE EXHAUSTED OF THE ACADEMY

Like Anaïs, both Lou and Sanja have expressed doubts about staying in academia, about continuing on the academic ladder. Sanja stated:

> It's not a safe system to work in. Everyone is following all these rules, collecting all these chips to stay in the game: you're role-playing through your career. In my head, in an ideal world, academia would be a place where people can focus on the important questions that they find meaningful. Instead, academia remains full of competition. Everyone keeps pretending they are excited about the thing that is more likely to be funded. That prevents people from focusing on what's important.

To me, their testimonies reveal something crucial about what motivates scientists into activism. It's not just the logic and the evidence with which the figure of the scientist is traditionally linked. Beyond the Gramscian logics of counter-hegemony, beyond the empirical evidence supporting scientist engagement, there is a deeper affective drive that pushes scientist-activists to participate in direct action.

There are feelings of rage, sadness, and disappointment, a yearning to break free from the navel-gazing of capitalist academia, to positively connect with the wider world around them. Indeed, social movement researcher Farzana Bashiri stresses that emotional engagement is a social driving force to scholar-activism.[33] This occurs not just in struggles strictly related to climate justice,[34] but also in connected and intersecting struggles against racism and marginalization.[35]

In his book *The Exhausted of the Earth*, Ajay Singh Chaudhary emphasizes the overlooked importance of a "generalized affectivity," a structure of feeling that propels political organization.[36] Specifically, he's referring to the affectivity of contemporary climate politics: a matrix of emotions—disgust, anger, and especially exhaustion—brought on by the global extractive circuit, which can motivate resistance and struggle. As academic capitalism pushes scientists to exhaustion, they also become enmeshed in this generalized affectivity, propelling them out of the lab (sometimes, as in the case of Anaïs, in the fullest sense of the expression). The same logics they experience in academia are writ large in the world of fossil capital—the same processes of extraction and competition, the same extreme concentration of power, and the same imperatives for infinite growth, infinite production, infinite exhaustion. These make plain and evident the cruelty of capitalism in its many other guises; they make it easier to understand other forms of injustice brought on by the same system. The bucket-wheel excavator is not just *out there* in the mine pit; it's also *in here* in the lab.

As they harness their emotions into action, Anaïs, Lou, and Sanja are asking themselves questions even nonactivist scientists are now forced to confront: Can universities become actors for good in the climate emergency? Can they transition away from the exploitative logics of capitalism? Or are they so full of pernicious constraints and incentives that they are not worth the effort? For now, the answer remains contested. Multiple actors are struggling to shape how higher education responds to a burning world. In the next chapter, we will see just how fierce this struggle has become: how fossil fuel corporations are infiltrating the hallways and classrooms of the university and how students and researchers are fighting back.

CHAPTER 6

The Enemy Within

"The only possible relationship to the university today is a criminal one."
—Stefano Harney and Fred Moten[1]

I put on a helmet and grab my construction tools. Today, I am part of an unusual building crew—made up of students, professors, and researchers. We are working on the central lawn of the Technical University of Denmark (DTU), one of the largest polytechnic institutions in Scandinavia. On top of the lawn, we are erecting a dog-training tunnel made of tarp. We've painted it black so that it resembles an oil pipeline.

We work under the worrying eyes of campus security officers, who promptly surround the scene, then tell us to stop. But we continue. Meanwhile, fellow academics display protest banners around the site, reading "DTU, stop taking oil money!"

It is November 16, 2022. Our mock pipeline is meant to evoke the very real pipelines that DTU is helping to build. The university promotes its commitment "to creating sustainable solutions that improve people's lives, create value for society, and promote a sustainable use of resources." It advertises itself as "a key player in the energy transition"[2] with a strong commitment to "creating sustainable solutions."[3] Yet, despite all these claims, in 2014, DTU received 1 billion

Danish kroner (approximately 140 million US dollars) from the fossil fuel multinational TotalEnergies. Now, in 2022, the university has just signed another million-kroner contract with the company. These agreements are not only enabling Total to carry out fossil fuel research on DTU's campus, but also helping the company improve its image in the eyes of the public.[4]

Total is the third-largest oil company in the world, responsible for roughly 1 percent of global CO_2 emissions between 1988 and 2015.[5] Even in the midst of climate breakdown, their website states they plan to continue producing massive amounts of oil and gas for decades to come.[6] Beyond its emissions, the company is infamous for destroying entire ecosystems in multiple countries around the world.[7] Its newest megaproject—the East Africa Crude Oil Pipeline (EACOP)—has been condemned by NGOs for causing freshwater contamination and wildlife habitat loss across its entire span, from Uganda to Tanzania. EACOP has also displaced thousands of small farmer communities in the area.[8] Once finished, the pipeline is expected to emit 379 million tons of greenhouse gases over its lifetime. Annually, this amounts roughly to the annual emissions of the entire country of Denmark, and over 25 times more than the current annual emissions of both its host countries.[9] The contradiction between DTU's sustainability pledges and Total's record could not be starker. For this reason, the university has been careful about broadcasting their fossil ties with the company. The initial contract between Total and DTU, for example, was made public upon signing, but it was later removed from the university website.[10]

On the day of our action, the DTU campus director hurried down to our construction site to see what all the fuss was about. We asked him if DTU's management would consider ending their deals with Total, to which he replied that DTU had too many projects together with the fossil fuel industry to make it profitable to disengage from them.[11] The event did not go unnoticed in Danish media.[12] A few days later, after repeated calls for explanation, DTU management felt compelled to defend their ties to the fossil fuel industry in a national newspaper.[13]

A DEAL WITH THE DEVIL

Our construction work would not have been possible without the help of another movement—Divest DTU: a group of students who had been demanding that their university stop investing in fossil fuel companies.[14] They were following in the footsteps of earlier divestment initiatives outside Denmark, which had been quite successful. For example, students at Harvard managed to get their university to divest from fossil fuels in 2021.[15] By the time of our action, DTU's management had stopped financing oil and gas companies, but they were still *receiving* money from them.

Why does it matter whether universities obtain money from the fossil fuel industry? A group of researchers recently showed that fossil money influences how academics talk about the climate emergency.[16] They performed an analysis of 1,706 reports published by academic centers working on energy and climate, across 26 universities. Three of these centers were primarily funded by fossil gas corporations. The study showed that the gas-funded centers tended to be much more favorable toward gas than toward renewable energies, in comparison to those centers without fossil funding. This influence was observed not just in academic outputs (scientific papers and reports), but also when scientists talked to the public on social media. Indeed, this is a strategy the fossil fuel industry is actively pursuing: in the leaked "victory memo" of the American Petroleum Institute, for example, fossil fuel executives describe their (successful) plans to create doubt about global warming by hiring scientists to talk on TV and in other media outlets.[17]

Fossil money is not just affecting the behavior of researchers; it also helps to improve the image of oil companies in the eyes of the broader public. Universities are places of immense societal prestige: any corporate actor that establishes partnerships with them can benefit from that prestige.[18] Oil companies advertise their contributions to higher learning institutions on their websites and public portals; this allows them to portray themselves as contributors to progress and welfare—a phenomenon known as "greenwashing."[19]

In the case of DTU, in addition to the contracts we were protesting, Total's money already served to fund an existing research institute. As advertised on their website, the Danish Offshore Technology Centre (or DTU Offshore, for short) has a mission to "develop research-based technology solutions for the offshore industry in the North Sea."[20] The "offshore industry" is here a euphemism for oil and gas extraction. DTU Offshore was originally called Centre for Oil and Gas before going through a rebranding effort in 2022.[21] Its founding document states that all projects within it must have "a clear line of sight to increased oil and gas recovery/production."[22] And in its steering committee, half of the seats are occupied by executives from Total.

INFILTRATING ACADEMIA

Our action at DTU drew inspiration from an earlier action by students and academics in Trinity College, Cambridge. Back in 2020, they dug up their campus lawn to call attention to Trinity's £9 million ($11 million) investments in oil and gas companies, and to the money they receive from them.[23] As with DTU, University of Cambridge's ties to fossil money are helping to launder the industry's name. Thanks to generous donations, the university is home to a "BP Institute," a "BP professorship," a "Shell professorship," and a branch of the Schlumberger offshore drilling company, also while claiming to be a "green university."[24]

DTU and Cambridge are not isolated cases. Over the past 3 decades, hundreds of higher learning institutions across Europe and North America have engaged in partnerships and sponsorship deals with the fossil fuel industry. In the United Kingdom, oil and gas companies funneled over £40 million ($49.5 million) into 44 universities across the country in 2022 alone. This has been in the form of research agreements, scholarships, grants, tuition, and consultancy fees.[25] The influence of the industry is so prevalent, some have called it a "colonization of academia."[26]

As in Denmark, the contributions often come from Total but also from other major corporations like Shell, BP, and Equinor. They all want a foothold on campus, and many universities are happy to roll out the red carpet for them. The University of Exeter is one of the most egregious examples, having recently signed a £14.7 million ($18.2 million) deal to fund research on carbon capture and storage, which in the words of a university spokesperson will "contribute to the global race to Net Zero."[27] By hosting such projects, universities become complicit in the strategy of the fossil fuel industry to create a smokescreen, to hide their polluting activities, as we saw in chapters 1 and 2.

The concept of "net zero" permeates many of these deals—a catchall phrase favored by oil and gas corporations to move attention away from decarbonization in the here and now. It also featured prominently in the communications campaign for the new contract between DTU and Total, which prompted our pipeline construction.[28] The word *net* does most of the heavy lifting: fossil fuel companies are not intending to downscale their emissions any time soon. The carbon emissions themselves are rarely put into question. Instead, net zero narratives are directed at some distant future—2030, 2040, 2050—and put the focus on investments in as-yet-undeveloped technologies (like negative emission technologies, NETs) that will, theoretically, allow countries to sequester more CO_2 than they emit.[29] Yet all the carbon removal equipment in operation today can only remove around 0.01 million metric tons of CO_2 a year[30]—far less than the 70 million metric tons a year that would be needed by 2030 to stay below the 1.5°C Paris Agreement threshold, as argued by the International Energy Agency (itself a carbon-capture-friendly institution).[31] As of 2024, the world has breached 1.5°C temperatures for the first time in recorded history;[32] even the Paris Agreement "threshold" itself—the very standard upon which carbon capture promises are evaluated and thoroughly fail—relies on more fantastical carbon capture promises. Fiction piles upon fiction. Ultimately, like the tobacco industry arguing that the cure for cancer is just around the corner, all of this is little more than a fairy tale told to the public in order to protect fossil capital and to kick the bucket farther down the road.

Meanwhile, universities like DTU continue conducting research into how to ramp up fossil fuel extraction. This adds even more weight to the burden that the "net" in *net zero* is supposed to carry. Rather than downscaling emissions, fossil-funded institutes are trying to find ways to increase them, pushing us as far from zero as they possibly can.

THE MANIPULATION OF POVERTY

The incursion of fossil capital is not restricted to Global North universities. In the Global South, academia is also plagued with financial ties to fossil capitalists. I spoke to a member of Scientist Rebellion in Nigeria—Ayo Fola—who described how fossil fuel companies there maintain a cultural hold over the oil-rich country:

> In Nigeria, to be a member of these companies is like a badge of social status, like something to be proud about. Politicians are very deferential to them, even though the oil and gas they extract from here is not for our people, it's for outside Nigeria. And these companies don't even pay taxes here.

As Ayo Fola explained, this cultural hold is often exercised through academic structures:

> One of the ways that fossil fuel companies push themselves as "good" is through scholarships. Over 80 percent of scholarships that students get in universities here are from fossil fuels. They also influence research because they give research-based scholarships. Literally they dominate scholarship schemes in Nigeria. The reason they do this is because they know that people with an education can more easily expose them: scientists, engineers, lawyers, people who can understand the science and the legal aspects of what they do. They don't go for arts and literature funding. They are giving money so people funded by them don't say too much about what they are really doing here.

Ayo Fola illustrates what climate justice researcher Dr. Roland Ngam calls the construction of a cultural hegemony in the service of the "expansion of the frontiers of accumulation."[33] In order to open new oil and gas projects, fossil fuel companies first need to win the hearts and minds of the local people whose land they want to invade. And in their effort to secure societal consent, students and scientists are the natural prey.

Fossil interests in Nigeria are centered on the Niger Delta. Since 1958, the Delta has been commercially exploited for oil and gas and has played host to over 5,000 oil spills, polluting the soil and the water.[34] One of the most outspoken figures in the struggle to rid the area of Big Oil was Kenule Saro-Wiwa—a writer, environmental activist, and member of the Ogoni people, who have traditionally inhabited the Delta for centuries. In 1995, Kenule—along with eight other Ogoni—was brutally murdered by the Nigerian government while they were waging a nonviolent resistance campaign against Shell. The murder provoked massive international outrage. Today, the Delta remains flooded with the toxic sludge of fossil fuel processing plants. Shell's oil leaks and gas flares have destroyed the livelihoods of local communities. The rivers—once vibrant sources of life—are now brown streams of death. People can no longer rely on them for fishing: most of the fish populations have collapsed. The water is so polluted it cannot be used for drinking or growing crops. Nevertheless, without access to clean water from other sources, many are still forced to drink from it, leading to high rates of diarrhea and other diseases in both adults and children.[35] Today, the Delta continues to be exploited by fossil fuel corporations, like Chevron, Total, and ExxonMobil.

Ayo Fola told me that academic influence via fossil scholarships is not a phenomenon unique to Nigeria. Total has sponsored several masters and PhD programs in other African universities, like Makerere University in Uganda, where it is building EACOP. And in the Global South, fossil capital has one more tool at their disposal: the manipulation of poverty. After decades of colonial and neocolonial abuses, poverty in both Nigeria and Uganda remains widespread. In Nigeria, for example, it is particularly concentrated in the Niger Delta—the main site of oil extraction. To greenwash their own poverty-generating activities, fossil fuel companies portray themselves as community benefactors to the Delta, as the ones who are repairing the damage they themselves continue to cause. Ayo Fola continues:

> Fossil fuel scholarships particularly target students from communities in certain vulnerable areas. They say they are "trying to pay back to communities," but that's bullshit. If they really wanted to pay back, if they wanted to support people in the Niger Delta, they wouldn't do what they are doing. Most of the young people in the Delta can't even afford universities because their means of livelihood have literally been destroyed. They are taking advantage of poverty. They use money to close people's eyes.

RESISTANCE FROM WITHIN

Fossil ties with academia are not going unchallenged. While we were working on DTU's lawn, a wave of protests was sweeping through Spain, Germany, the United Kingdom, and the Netherlands. It was largely made up of university students—members of a radical collective called End Fossil: Occupy!, a student network whose shared goal is to "end the fossil fuel era." Their tactics included occupying schools, universities, and other institutions, sometimes for days or even weeks.[36] Among End Fossil's most prominent targets was the prestigious Erasmus University in Rotterdam, where students set up a "climate camp" on February 7 and 8, 2023. A local chapter of End Fossil called OccupyEUR was demanding that Erasmus end all of its fossil ties, particularly with the multinational Shell. The occupation occurred after repeated requests to Erasmus's management, who had refused to heed the students' calls.

Back in 2012, Shell had paid Erasmus hundreds of thousands of dollars in exchange for a contract that allowed the oil company to have a say in the design of the university's curriculum, as well as "the profile of the students who attend."[37] The corporation effectively had a VIP pass to shape what students were being taught. The contract was shrouded in secrecy, and it was only uncovered after an extensive investigation by one of Rotterdam's own researchers in 2017.[38] As revealed by internal company emails made public via a freedom-of-information request, this was part of Shell's strategic effort to target young people and academics. The multinational was seeking to position itself as a "low carbon leader," even as it spent billions of dollars in efforts to drill in the Arctic.[39] The university was honoring their agreement by regularly inviting Shell to career fairs and involving it in widely publicized "sustainability" research projects.[40] To top things off, the university's trust fund was still investing in fossil fuel ventures. The students at Erasmus were outraged: they would not leave willingly until their university put an end to its oily dealings.

The management's immediate response was to call the police. One by one, the students were violently arrested by officers in riot gear.[41] Erasmus then tried to deny some of the accusations advanced by the students. An official press release by the administration claimed they were not investing in fossil fuel corporations at all—a statement later shown to be false, and then deleted from their website.[42]

Dr. Willem Schinkel—a professor of sociology at Erasmus—made a postmortem analysis of the management's reaction to the occupation:

> This response testifies to an utter incomprehension of campus protest, and to a kind of housekeeping reflex, a neurosis of security and hygiene. When students were unwilling to, on day one, dilute their protest to a "dialogue" on the administrators' terms, the administrators' response was, entirely in keeping with the corporate identity of the university: get the fuck out of hEUR [Erasmus University Rotterdam] with your attempts to make of this place something more than a factory for credentialization and a lobby lounge for suits and ties intent on doing what their daddies did before them: cashing on the planetary plunder called capitalism.[43]

In his analysis, Schinkel stresses how leadership sought to externalize its own students when they resorted to protest, to frame them as outsiders who were no longer part of Erasmus's academic community:

> Apparently, administrators fail to recognize protest unless it is flattened to "having a certain opinion" and expressing it in a format they determine (a "dialogue"). And with a historical and political-theoretical amateurism that is almost touching, they believe a protest is something that doesn't

disturb anything. . . . Whatever isn't recognized as "academic community" in the anti-intellectual and ahistorical narrow-mindedness of the administrative frames is repressed by police violence.

End Fossil is not the only group resisting the fossil colonization of academia. A sister movement—called Fossil Free Research—has recently drawn mass support from faculty and students in the United States and the United Kingdom.[44] In March 2022, they published an open letter signed by almost 500 scientists, including members of Scientist Rebellion like Dr. Peter Kalmus, calling on universities to stop accepting fossil money. Chapters of Fossil Free Research then started multiplying and began organizing campus sit-ins. The protests have been fueled by the newly found rage students experience as they learn about their universities' links to fossil fuels. As Ilana Cohen—lead organizer of the movement—explained in an interview: "We're drawing attention to how the fossil fuel industry continues to infiltrate prestigious academic institutions, mooch off their credibility, and even exert influence over the production of knowledge crucial to shaping climate policy."[45]

End Fossil, Scientist Rebellion, and Fossil Free Research are relatively young movements, but their efforts are already paying off. After staging a sit-in at Cambridge's fossil-funded BP Institute, Fossil Free Research got the university to remove "BP" from its name and also to hold a vote to ban fossil money from research. Under sustained pressure from within, departments in other universities have followed suit, enacting bans on funding schemes linked to oil and gas corporations.[46] In the Netherlands, under sustained pressure from Scientist Rebellion and Fossil Free Research, the Vrije Universiteit Amsterdam enacted the first ban on research partnerships with fossil fuel companies that do not show "demonstrable commitments" to the Paris Agreement, while the University of Amsterdam has placed a moratorium on collaborations with oil and gas multinationals like Shell.[47] In Spain, soon after the events at Erasmus University, another days-long student occupation at the University of Barcelona led to the enactment of mandatory campus-wide climate science education, for the first time in history.[48] Today, climate curricula and fossil money bans are being discussed in hundreds of higher education institutions. Their students are forcing them to confront their complicity and responsibility in the face of climate breakdown.[49]

OILING THE WHEELS OF PUBLISHING

The influence of fossil fuel corporations on academia extends beyond the narrow confines of the university. Many external industries with which academics work also service Big Oil, Gas, and Coal. Chief among these is the publishing industry. We saw earlier how this industry has manipulated academics into depending on commercial journals, through a strategy of manufactured prestige built over decades of marketing. But academics are not the only source of revenue for big publishing companies. These corporations are also engaged in backroom deals with fossil fuel multinationals. Top publishers like Wiley and Taylor & Francis, for example, have signed partnerships with oil giants and produce several journals designed to help the industry find more fossil fuels.[50]

Among the world's biggest publishers, one stands out from the rest. The Dutch company Elsevier holds a portfolio of over 2,800 academic journals. These include popular brands like *The Lancet* and *Cell*—both world-famous in the health and biomedical fields. Elsevier is also a highly profitable corporation: its products rake in close to a billion dollars each year, and its profit margins are among the highest in the industry. The company has made numerous public statements about its dedication to sustainability. It has an entire website dedicated to advertising its commitment to the Paris Agreement and the UN's sustainability goals.[51] This includes what an executive calls an "ambitious Climate Action program to accelerate our efforts to decarbonize our business." Elsevier has also boarded the net zero wagon, pledging to "become net zero for all emissions by 2040 at the latest."[52]

Kip Lyall is a former employee of Cell Press—a subsidiary of Elsevier. While working there, he started questioning whether the company was living up to its word, whether its operations were actually as sustainable as they were claimed to be. Yet the more questions he asked, the more he realized this could not be farther from the truth. In his investigations, Kip found that Elsevier has been giving the fossil fuel industry all kinds of assistance in order to expand their operations, way beyond what is compatible with the Paris Agreement, their net zero pledges, or any basic notion of sustainability. He found that Elsevier provides information about areas yet to be exploited for oil and gas extraction, as well as new technologies needed to dig up oil more efficiently, in difficult-to-access locations like the Arctic and the Canadian tar sands.[53] Elsevier's online tools such as Geofacets and Knovel help fossil fuel companies to efficiently amalgamate data and find promising sites for exploration and extraction.[54] Indeed, the motto of Geofacets used to be that, thanks to it, "exploration has never been more seamless."[55]

Kip realized that Elsevier's ties with fossil fuels run even deeper than publishing and data services. Its parent company—the multinational RELX—is one of the biggest publicly listed information conglomerates in the world. For years, RELX's political action committee, REPAC, has been providing support to climate denialist politicians in the United States, preventing carbon mitigation bills from reaching a vote in Congress.[56] Additionally, RELX's RX division has hosted numerous exhibitions for the fossil fuel industry, including the Queensland Mining and Engineering Exhibition and Asia Pacific's International Mining Exhibition.[57] As he opened Pandora's box, Kip tried to reach out to various corporate managers for an explanation. He was trying to understand how this could be, how a company that boasts so much about being "committed to climate action" could be contributing in so many ways to climate breakdown.[58] Kip shared with me the Kafkaesque nightmare of delay, denial, and gaslighting that followed:

> I had reviewed Elsevier and RELX documents and was shocked to see the extent of the oil & gas services they had. The R&D, all the books they publish, the exhibitions serving oil & gas—it was overwhelming and contrasted with their many pledges around the Paris Agreement and the UN

principles on human rights. While I was still employed, I had created a lengthy report about all I had found. In April 2021, I sent the report to my manager, and to her manager, and individually emailed it to 120 people at Cell Press. I asked them what they thought about this. I also filed an official ethics complaint with HR.

Kip did not stop there. He contacted top and middle managers across the corporation in his quest to get someone to answer for the company's policies:

It appeared that a big management technique was evasion, denial, and disengagement. Managers tried to deny the existence of Elsevier's concrete sustainability pledges and statements. I was subsequently told in sum and substance that the company's pledge is just that they "care about these issues." They put me in meetings with managers who seemed not to know much about the Paris Agreement or to have even read the report I had prepared. A group of employees asked RELX to provide a time frame for when the company will disengage from supporting fossil fuel expansion, and the company never provided an answer. I reached out to the ethics department after never hearing from a credentialed investigator for my ethics complaint, and they said they had determined there were "no code violations." When pressed for answers, they seemed to assert it was not their place to comment on the company's position on climate change.

It was an exhausting quest. And Kip struggled with confronting his employers, who were supporting the fossil machine on a daily basis:

We're used to the bad guys being these faceless corporations. The heart of my difficulties was interacting with the human beings behind them, those in charge of protecting revenue at the cost of everything else. Throughout these interactions, it became very clear to me that there were no adults in the room.

All of his inquiries culminated with Kip being fired from Elsevier. The company simply found it too inconvenient to have someone prying so relentlessly into its dealings with the fossil fuel industry.

But he did not leave without a fight. Kip reached out to academics from Fossil Free Research, Scientist Rebellion, and Scientists for Extinction Rebellion—the groups that already had experience in the fight to rid academia of its fossil ties: "I realized the company is not going to do anything without external pressure, which could come from their customers: the scientist that they serve." These groups launched a campaign called Stop Elsevier in early 2022 to spread the word about things Kip had found and to get the company to stop aiding the fossil fuel industry.

So far, the campaign has included boycotts—including refusals to publish, edit, or review in Elsevier journals—as well as more direct forms of disruption. In the United Kingdom, the group Scientists for XR organized a rally at the entrance to RELX's shareholder meeting.[59] One of the collective's members—Dr. Caroline Vincent—bought a RELX share and interrupted the meeting to ask the company's leadership why they continued to help the fossil industry drill for new oil and gas fields. The scientist-activists in this campaign have also found common ground with earlier academic initiatives targeting Elsevier for its deals with the weapons industry[60] and its exploitative publishing practices.[61] Most recently, a large coalition of academics and climate NGOs initiated a legal grievance mechanism at the UN Human Rights Council, calling out the company for the risks it poses to human rights as a direct consequence of its support for fossil fuel corporations.[62]

TOPPLING THE IVORY TOWER FROM BELOW

Why are universities and publishing organizations so liable to be captured by fossil interests? One of the features that seemingly permeates all these institutions is what Kip describes as a generalized "tunnel vision": a disseminated and enforced refusal to confront the consequences of one's activities. This tunnel vision drives those in power—think the DTU director, the heads of the Nigerian universities, or the various bureaucrats at Elsevier—to avoid engaging with the role their institutions play in fossil fuel emissions, as long as the money keeps flowing in. It also drives the public faces—the HR reps and the sustainability "stewards"—to avoid confronting difficult questions within their own organizations.

Another feature of these institutions is a co-opted vocabulary: concepts like "sustainability," "green," and "net zero" have all been repurposed into new meanings, or voided of any meaning at all. Instead of organizations adapting to the dangerous reality at hand, it's the words themselves that are being forced to adapt. In political theory, this phenomenon has been called *hegemony through neutralization*: the meaning of once-powerful calls for change is appropriated, reinterpreted, and rendered powerless by actors intent on continuing with business as usual.[63] As Kip points out, "to help avoid responsibility, companies find ways to map existing activities they are already performing to their sustainability goals, rather than changing activities with negative impacts in accordance with those goals. That way, they can say they are doing what they are supposed to be doing."

Yet, tunnel vision and co-optation are symptoms of a deeper issue. As long as companies stand to profit from their dealings, they are likely to continue doing so. To the extent that Wiley, Taylor & Francis, and Elsevier are profit-seeking corporations, they will continue acting as such. We should not be surprised by this; it is what corporations are meant to do: secure profits for their shareholders. And to the extent that university managers like those at DTU, Makerere, Cambridge, and Erasmus also see themselves as running corporate entities—rather than institutions of learning and inquiry—they will also think twice before ending their ties with Total, Shell, and BP.

This, then, is the crux of the issue. Appealing to these institutions' good nature, like asking their managers to be more sustainable, is unlikely to be a winning strategy. Instead, activists are learning to find their weak points and to hit them where it hurts most. Student occupations, interruptions of shareholder meetings, and unauthorized construction projects on campus lawns are ultimately PR stunts, because it is PR that these organizations rely on to continue with their dealings. Business with Big Oil, Gas, and Coal depends on a veneer of normalcy with the rest of society, and it is this veneer that scientist-activists are trying to bring down.

Hierarchy within these organizations also plays an important role: as Kip experienced over and over again, the tunnel vision and co-optation appear to be enforced from above. Likewise, in his postmortem analysis of the student occupation at Erasmus University,[64] Willem Schinkel calls out the university for its "extremely hierarchical bureaucracy, based largely on autocratic government." Yet all of the protests I have talked about in this chapter were driven by those in the lowest rungs of the hierarchy—students and staff, often with precarious working contracts—the ones with the least power to influence policies through "regular" channels. Fossil agreements are made by a powerful few, sitting at the top of the ivory tower, and they are often done in secret.

Thus, Schinkel emphasizes that the student occupation at his university was not just about the ties to oil companies, but about the lack of academic freedom and democratic practices:

> The fact that their protest was thus also a fundamental defense of academic freedom is entirely lost on the bureaucratic squares who believe the university is first and foremost a "brand." Yet that protest can be of peripheral interest to no one who thinks academic freedom matters. Next time, look up from your tenth paper this year, walk out on your meeting.[65]

Ultimately, to break with business as usual, academics are finding they need to break with science as usual, to build counter-hegemony: to "walk out," to question how power is distributed, how decisions are made, and who is allowed to say what. Increasingly, scientist-activists are leading by example, disobeying top-down mandates dictating how their science is supposed to be practiced.

CHAPTER 7

The Road Less Traveled

"The skillful traveler leaves no traces of his wheels or footsteps."
—Lao Tse[1]

"Though they carry nothing forth with them, yet in all their journey they lack nothing. For wheresoever they come, they be at home."
—Thomas More[2]

"Fly like you own it."
—Promotional material—Symphony Private Jets[3]

Dr. Gianluca Grimalda speaks with a deep, gentle voice. He has a dignified demeanor, kind eyes, and a bristly gray beard—almost exuding wisdom. He is a social psychologist. Through his research, he seeks to understand how people organize into groups, how they interact within and across communities, and how they respond to their environment. He is especially interested in the social psychology of climate change: in how societies are responding to increasing weather catastrophes.

One thing sets Gianluca apart from his colleagues. For the last 10 years, he has been working on the impact of increasing storms and sea level rises in the island of Bougainville, in Papua New Guinea.

However, Gianluca lives in Kiel, in northern Germany, half a world away from his field of study. The typical solution to this conundrum would be to take a 32-hour flight, carry out surveys and interviews on-site, then fly home again after a few weeks of research. Yet, Gianluca refuses to board a plane. He has been a "conscientious objector to flying" for more than 10 years.[4] He is wary of the emissions that the flights to Papua and back would generate: a bit over 10 metric tons, or approximately what the average European emits in 2 years, and what the average Papuan emits in a decade. In conversation with me, he explained why he made this commitment:

> I am inspired by the group Scientists for Global Responsibility, who in 1997, traveled overland from UK to Kyoto to attend the 3rd UN Climate Change Conference of the Parties (COP3). They knew that traveling by plane produced 8 times more emissions than by land, so I thought I should do the same. For me, it is a personal commitment: I want to do as much as I can to limit my emissions.

For his latest research project, Gianluca chose to take the long way to Bougainville. In early 2023, he embarked on a 42,000 km land and sea journey, using buses, trains, and the good will of people kind enough to haul him along the way. In total, he traversed 17 countries, across three continents, following in the footsteps of the great transcontinental travelers of centuries past.

LIVE-TWEETING ON THE SILK ROAD

Told through Twitter posts, Gianluca's journey reads like the modern-day version of a nineteenth-century naturalist journal.[5] The slow pace forced him to come face-to-face with the difficult realities of the many people he met along the way. On the sea crossing from Greece to Turkey, he encountered refugees from the Syrian civil war, which took the lives of over 600,000 people[6] and forced millions more out of the country.[7] Later, the border checkpoint into Iran confronted him with dozens of hungry preteens hauling luggage for less than a fraction of a US dollar. In Tehran, at the height of the protests in defiance of compulsory hijabs, he met women on the street who were breaking the law and leaving their hair uncovered—a form of civil disobedience that could cost them their lives.[8]

Many of the stops Gianluca made are already testimony to a dangerously warming world. Once in Iran, he passed through the Lut Desert, where the hottest land surface temperature in historical record, 70°C, was recorded in 2005. Farther southeast, in the Balochistan region, he saw the remnants of houses, bridges, and roads destroyed from mass flooding in 2022.

> I traveled through some of the areas most affected by major flooding in Pakistan. Some people had been evacuated many months ago and had not gone back to their houses. At the same time, it was strange to think flooding had affected these areas so recently, as all the rivers were completely dry. Yet in the summers, when it gets as hot as 50 degrees Celsius, the heat and moisture accumulate and produce torrential rain, destroying entire villages.

The dried-up rivers were already affecting the region's agriculture. Balochistan is experiencing droughts that used to happen only once every few hundred years but are now a common occurrence. Thus, food is becoming harder and harder to grow. On top of this, the people Gianluca met there are facing not only climate disasters, but the threat of violence from conflicts between armed separatists and local governments.

Eventually, Gianluca managed to cross into India. There, on the train to Kolkata, his live tweets highlight what remains hidden for those who choose to fly:

> Looking at slums through the train window is a punch in the stomach. People live in shacks close to or among garbage dumps. [I] feel ashamed of being born in incomparably better conditions. With flying, this reality is hidden away from your eyes. With slow travel, one gains awareness.[9]

Despite the long journey, there was one constant that permeated all his experiences:

> The amount of completely disinterested help that I received was just incredible. Along the way, many people welcomed me as if I was a member of their family. Twice people paid for my taxi, when I had run out of

cash. On a 3-day bus ride along Sumatra, one guy treated me like a brother. On the last day, he gave me a little bottle with a flower inside, and he told me, "This is my present for you; this flower only grows atop a mountain in Java island." He cared for me and he asked for nothing in return. Another time, I was traveling by train with an Indian couple with a small daughter, who had just been kicked out from Italy because their working permit had expired. They were 10 times poorer than me, but as soon as they met me, they offered me food! They had every reason to be angry against my home country, but they still shared food with an Italian.

Gianluca speculates this might be due to the historical vulnerabilities of travelers in ancient times:

Some people told me that there is a habit or social norm to really treat travelers as if they were members of one's family. In the past, travelers and pilgrims were completely vulnerable when arriving in a new town or settlement. They were subject to the whims of local rulers, who could even dictate they be put to death. So, my guess is that, for this reason, social norms of reciprocity emerged. You are expected to welcome travelers that arrive at your doorstep, because someday one could also be the traveler needing help.

In large part thanks to this kindness of strangers, Gianluca eventually made it to his research destination.[10] All in all, Gianluca's trip from Germany to Bougainville took 72 days. On two occasions, due to land border closures and armed conflicts, he had no option but to take flights, across Burma and Indonesia.

For an academic, Gianluca's travel patterns are quite far from the norm. Few scientists would have chosen to spend months traveling by land, especially if their field site was in a distant continent. But Gianluca's abnegation of flying does not make him much more different than the majority of people around the world, who have never even boarded a plane. It is estimated that only between 5 and 20 percent of the global population has ever taken a flight.[11] Most people cannot afford to do so, or they are not even permitted due to migration policies.[12] So why is it that low-carbon travel is so foreign to academics?

ACADEMIC HYPERMOBILITY

Academics fly a lot. They fly for field trips and research expeditions. They fly to present their work at conferences. They fly to meet other academics and potential collaborators, to interview for new positions, or to get training in new research areas. In fact, academics fly much, much more than the average person.[13] A 2009 survey showed that ecologists and sustainability scientists emitted on average more than 9 tons of CO_2 per year.[14] Another survey in a Norwegian atmospheric research institute revealed that its members annually emitted up to 2.4 tons on average, just in air travel.[15] To put these numbers into perspective, the average French person emits 4.6 tons of CO_2 *from all activities* (including travel) in an entire year. The contrast is even starker when looking at average emissions in the Global South: an average Brazilian emits a total of 2.2 tons per year, and an average Ugandan emits just a tenth of a ton.[16]

Frequent flying is seen as a mark of prestige in academia, as something to aspire to.[17] And for those who can fly, it comes to define one's identity—the Swedish tropical ecologist and her annual field season in Martinique or the dual-appointment Canadian historian and his summer course in Beijing. Who would they be without long-distance travel? Andreas Malm considers this an ideological affiliation, by which one's idea of oneself is built by the very act of flying:

> Whether a party animal or a progressive academic, you need to take that flight to maintain your subjectivity and be the person you are . . . you cannot imagine not flying: the [fossil fuel] has constituted the subject, who cannot see himself outside of it and who rarely reflects upon, let alone articulates, the ideological affiliation. It is just there, in the veins of material life.[18]

The incentives to fly are both ideological and structural. Many research organizations reward academics for moving across distant countries, via funding schemes that foster "mobility" and "internationalization."[19] Students are encouraged to take postdoc stints or lectureships abroad—*perhaps then you might be able to come back and find a job here*. It is as if the fact of flying, of going away, supersedes the actual research or teaching we might practice wherever we happen to land.

The incentives to fly also emerge through policies that disincentivize trains or buses. Ground travel takes time, which is scarce for precariously employed academics or for those who must take care of children. And thanks to massive government subsidies, taking a plane is often cheaper than going by train. So even when universities might extol the virtues of sustainable travel, they rarely provide the time or resources for their employees to emit less: *Going to a conference abroad? Be sure to be back next week.* The incentives are also in the form of symbols. My university's default icon for travel is a plane; I am reminded of the possibility to fly every time I record my train rides on their reimbursement portal.

But not all academics fly in equal amounts. Ironically, hypermobility is prevalent among climate scientists, whose flight emissions tend to be larger than those of other scientists.[20] Differences in emissions are also associated with employment categories. Echoing the carbon inequality between the global rich and poor, airplane emissions are much higher for those at the top than those at the bottom of the academic ladder. In a 2018 Canadian survey, 50 percent of flight emissions were attributed to approximately 10 percent of academics.[21] And in a major institute in Lausanne, the top 10 percent emitters accounted for 60 percent of all flight emissions between 2014 and 2016.[22] Typically, it is those with more secure positions and higher salaries—full professors, managers, and research group leaders in the Global North—who are responsible for most flying emissions, in comparison with students, employees with temporary contracts, or researchers from the Global South.[23] For example, the Lausanne survey found that an average professor emitted 10 times as much as an average doctoral student. In Zurich, the emissions from 1 professor were found to be equal to those of 8 doctoral students, 22 research technicians, or 78 bachelor students![24] While Swiss institutes tend to keep extensive records of who flies where, these patterns are not unique to that country: in a large survey of flights at the University of Montreal, professors were found to emit on average 3 times as much as doctoral students.[25]

The idea that flying is necessary for one's success has thus become ubiquitous.[26] One could even add a fourth entry in our list of academic commandments:

Thou shalt fly, or else perish.

Like those maxims pushing scientists to publish fast and strong, or to stay away from politics, this maxim is a fairly recent one. Flying was not seen as an essential part of being an academic before the advent of cheap commercial aviation. Yet, even as this imperative looms large, there is little evidence that flying helps much in academic advancement.[27] Many of the performance metrics that scientists have been co-opted into using (as we saw in chapter 5) don't even seem to be that much affected by the amount a person flies, after controlling for seniority.[28] Instead, the predominance of flying among senior academics suggests that it may be an elite luxury of the successful, rather than an actual contributor to success: scientists who have already achieved high status simply end up hopping on more planes.[29]

MAKE THEM PAY

While academics fly a lot, they are not the heaviest contributors to flight emissions. The top polluters are by far society's wealthy elites: the top 1 percent is responsible for about half of the CO_2 from global aviation.[30] If you're looking for the starkest example of climate injustice, look no further than your nearest airport.

Scientist-activists are beginning to focus on the emissions of the ultraelites. Just months before his journey to Papua, Gianluca paid a visit to Milan Linate Prime—one of the busiest private jet airports in Italy. The reason for his visit was not to board a plane, but to prevent one from departing. He sat down on the road and chained himself to the airport gate. Along with nearly 200 other activists, including 123 scientists, Gianluca was part of a 16-airport blockade, which took place in mid-November 2022, across 11 countries in Europe, North America, and Australia. For his actions, Gianluca was arrested and

FIGURE 13. Occupation of the private jet tarmac at Schiphol airport in Amsterdam by members of Scientist Rebellion and Extinction Rebellion Netherlands. Credit: Sjoerd van Beelen.

shoved into the back of a police car. As he was being apprehended, he explained why he was resorting to these measures: "I'm a researcher and I must do this because otherwise they will go on with business as usual. We face a serious risk of extinction."[31]

The actions—including blockades in London, Berlin, Melbourne, and Los Angeles—were undertaken in coordination, as part of the global Make Them Pay campaign. The campaign was organized by a large network of over 200 organizations, aptly called Stay Grounded, which included Scientist Rebellion as one of its main partners. This was a new wave of rebellion: now that they were organized internationally, scientist-activists were not just acting on the same day, but *toward the same demands*. In particular, the Make Them Pay campaign was pushing governments to (1) ban private jets, (2) tax frequent fliers, and (3) make the wealthiest in society pay for climate damage already done.

The demands were chosen strategically, not just to decrease aviation emissions, but to make a link between these emissions and the wealthy few. It was a call not just for climate justice, but for social and economic justice as well. Private jets are 5 to 14 times more polluting than commercial planes per passenger, and 50 times more polluting than trains. For example, a person on a 4-hour private flight emits as much CO_2 as the average European person emits in an entire year.[32]

The group of people who own private jets is tiny and extremely wealthy: they represent just a fraction of a percent of the global population, and their median net worth is $190 million. They largely come from the banking, finance, and real estate industries and are overwhelmingly male and over the age of 50.[33] But the effects of the ultrarich on our climate are colossal: as a recent Oxfam report shows, the crop losses caused by the emissions from the consumption, travel, and investment decisions of the top 1 percent could have provided enough food to feed 14.5 million people a year in the past 23 years.[34] This number will rise to 46 million annually in the next 3 decades. The Make Them Pay campaign wanted to direct attention to them: the jet-setting oligarchy.

Since Gianluca's arrest, Make Them Pay has scored several victories. On April 4, 2023, after continued pressure and repeated occupations, the CEO of Amsterdam's Schiphol airport announced they were going to ban all private jets from their tarmacs.[35] Soon after, the Eindhoven airport, also in the Netherlands, announced a private jet ban starting in 2026.[36]

THE PRICE FOR NOT FLYING

Beyond their concrete achievements, the airport protests are helping to denormalize hypermobility in academia. It becomes harder for academics to boast about their frequent-flier miles when coworkers are arrested for protesting against frequent fliers themselves. This negative reinforcement effect complements the positive messaging from other grassroots movements, who are making trains and other travel alternatives more socially appealing in academia. The Flying Less initiative, for example, facilitates and encourages academic work that entirely avoids or minimizes flying. Here, social media plays a big role: academics share photos and videos of their train stops along their long-distance travel routes, signaling to others that "slow travel" is the way to go. Thus, scientist-activism in the streets flows back into the university, shaping which types of activities are socially acceptable, and which are frowned upon.

As they slow down, many academics are also discovering (or rediscovering) the pleasures of traveling slowly: more space to sit, better scenery, and less jet lag. Yet, slow travel is not all sunshine and roses. Most universities still make it difficult for academics to travel by train or bus, encouraging fast air travel instead. As mentioned earlier, this is not done explicitly, but through hidden structural barriers that favor the cheapest travel option or the one that minimizes travel time.

Sometimes, the institutional drive to travel fast is made manifest in more obvious ways. In 2023, after 6 months of conducting his research in Papua, it was time for Gianluca to return home to Germany. This time, however, he had made a promise to the people he had met in Papua—many of whom were being directly affected by climate breakdown. He told them his return journey would be absolutely flight-free, without exception. This meant he would have to find a way to hop across Oceania and Indonesia to get to the Eurasian mainland—a difficult stretch of travel he had made via air on his way to Papua.

Gianluca was successful; he found his way to mainland Southeast Asia using a combination of dingy barges and catamarans. Eventually, he made it back to Germany entirely by land and sea. Overall, he managed to avoid 7.6 tons of CO_2 that he would have otherwise emitted had he flown the whole way, to Papua and back. Yet, his return journey would cost him more than the many train, bus, and ferry tickets he had to buy along the way. On October 11, Gianluca received an email from his employer, the Kiel Institute for the World Economy: he had just been fired.

During his stay in Papua, the Kiel Institute was pushing him to take a plane back. They could not imagine that a researcher could take so long to travel for carrying out survey work. Soon before he was meant to start his return trip, his managers in Kiel told him he needed to be back in Germany next week, or else he would be sacked:

> They had approved a 7-week journey for the return. The key problem was that my fieldwork had been delayed by about 7 weeks. So they said that my permit had expired. But there was nothing I had to do back in Kiel. I said to them that I was prepared to go back to Germany with unpaid salary. I told them they could choose how many months of unpaid leave they wanted me to get, and I would let them do it. And they still said, no, we want you here now.

This was the price Gianluca had to pay for staying grounded. Yet, he soon realized it was not such a high price after all. His firing garnered widespread attention from the press, even more so than his arrest at the airport in Milan. He has now been offered book deals and a chance to star in a documentary about his low-carbon travel experiences.[37] As he puts it, "it's crazy one has to lose one's job to get this type of attention, but now I am able to reach a lot more people and generate more awareness about climate change than I was able to do before."

SCIENTIFIC PRACTICE AS A POLITICAL ACT

In a recent panel debate, I found myself arguing that scientist-activism is not categorically different from other methods academics use to influence public policy. The moderator disagreed with me. "Surely," he argued, "there are 'apolitical' ways of doing science. For example, there is quite a big difference between a climate scientist measuring ice thickness in Greenland and what Rasmus Paludan does."

I was struck by the comparison. Paludan is a fascist agitator, famous for his racist anti-immigrant agenda. He often appears on Danish and Swedish news, filmed while burning Qurans to provoke retaliatory violence from the Muslim community. What did the moderator imply by his statement? Was he suggesting scientist-activists blocking airports, like Gianluca, were on equal footing with incendiary far-right provocateurs? I took a breath and started dissecting the question. In a sense, yes, both Rasmus Paludan and scientist-activists are engaging in politics. In very different kinds of politics, surely, with very different goals, but both are trying to shape how society is organized. But let's leave Paludan aside and focus on the other example the moderator used: the lone scientist measuring ice in Greenland who is, supposedly, an apolitical actor. Politics—from the Greek *polis*, "city-state"—is about engaging in collective debates and disputes about how to organize society, and ultimately about the power relations that define how we cohabit with each other. When scientists generate knowledge, and when they share that knowledge with society, they are contributing to those debates and disputes; they enrich them and help to shape them. And so, the moderator's archetypal apolitical scientist—diligently taking measurements in Greenland—is (consciously or not) making just as much of a political statement as the other two. Such a scientist is shaping society, not just by the information being collected, but by the way of collecting it. Who is this scientist? Is it a *he*? How is his research funded? Is this scientist European? And if so, why is he in Greenland? How did he get there? Did he fly? Or did he, like Gianluca, take the road less traveled?

These are all political questions, related to the history of the West—a history mired by colonial violence and by the complicity or involvement of academia in "educating" colonized territories in Asia, Africa, and the Americas.[38] For example, long-standing partnerships between Danish and Greenlandic institutions allow Danish scientists to have privileged research access to the Greenlandic territory today, to carry out experiments or perform fieldwork there. In large part, this is due to historical colonial rule in the region. Ice-measuring scientists do not emerge out of thin air; they are situated in a political sphere, and their actions influence that sphere.

Beyond *how* our hypothetical person would be measuring ice in Greenland, one might also ask a different question: *why is he doing so, and why is he not doing something else?* In times of a cataclysmic climate emergency, are more measurements of ice thickness really what society needs? This is far from being a thought experiment. Indeed, in a recent article, a team of sustainability scientists asked themselves this exact question, evincing the political nature of the work they are meant to do:

What next for climate change science? Do researchers continue to provide ever more data, novel collaborations and forms of outreach, and participate in more marches and petitions, hoping that governments will respond to the scientific consensus, mounting impacts and growing urgency for action? . . . The tragedy is continuing research when the problem is political, diverting attention away from where the problem truly lies, and being gaslighted into crafting new scientific institutions, strategies, collaborations and methodologies.[39]

Based on this critical self-reflection, the article's authors took an explicitly political stance. They called for a complete "moratorium" on all climate research—including the publication of IPCC assessments—until global leaders take seriously the information they have been providing for decades and "mobilize coordinated action."

Other climate scientists are not stopping their work, but they are reconsidering and refocusing how their work is practiced. Lou—who we met earlier in chapter 5—was, like our hypothetical scientist, also a person studying ice in Greenland. But her experiences there and her increased involvement in climate activism made her reflect on the environmental effects of her research:

It's a weird thing to be jumping between planes and helicopters, emitting huge amounts of carbon emissions just to transfer us and the five tons of food we need to be there, in Greenland, to study how a glacier is dying. The second year I was there, our solar panels stopped working, and we had to use diesel generators. I woke up every day and saw these dark fumes pumped into this pure, crystal environment. It was very dystopian. And the generators kept breaking, so we had to have a helicopter come every 5 days to bring a new generator.

After two rounds of research in Greenland, Lou realized she had had enough. She started working with a colleague—Adrien, also a climate scientist and an activist—on an alternative way of conducting her research:

Two years into my PhD, I wanted to do things differently. After the second expedition to Greenland, the cognitive dissonance related to the carbon cost and colonial legacy of my research struck me. Sitting with my internal contradictions impacted me more than I thought, and I wanted to channel these negative emotions into a change. Adrien and I decided to develop a new research project on a glacier in Norway. We are based in Denmark, so we took the bus to Oslo, and then a train to the northern town of Finse, home to one of the largest glaciers in Norway. It was more laborious work: there were no planes, cars, or helicopter to carry our stuff—we had to do everything ourselves—but it was also a much more grounded experience. We could try to understand what was happening to this glacier with a lot more humility than how we had behaved in Greenland. After this, I learned that I don't want to hop on a helicopter ever again in my life.

Freed from the painstaking preparations of fieldwork in Greenland, Lou found she had more time to dedicate to activism. It was an emotional decision to abandon her original field site, but it was also a decision charged with political meaning: she was choosing to stop contributing to the very problem she was studying.

Gianluca's decision to stay grounded on his return from Papua was also a political act—one that shed light on academic pressures to travel fast and frequently, and thereby to pollute. It also highlighted Gianluca's right *not to fly*, to live and move sustainably, a right that is often overlooked in debates about academic freedom. Indeed, the decision by his institute to fire him stirred a colossal response from the wider scientific community: over 9,000 people, including well-known climate scientists, signed a petition for him to be reinstated.[40]

It is difficult to think how any scientific act can be apolitical. Indeed, an apolitical activity is an activity that—under the reigning value system—is so well sanctioned that it does not stir any discussion or debate, an activity that does not raise any eyebrows. Yet, by its very nature, science is meant to raise eyebrows: to question, to find out, and to change how people think. To be apolitical is to avoid posing difficult questions. This is the very opposite of what scientists are supposed to do.

Sometimes, the "political" label indicates excess. In science, one should not be seen as being *too* political, as if there were some cutoff over which politics is somehow *too present*. Yet, the amount of political meaning of an act is entirely dependent on the context in which that act is performed. This applies in science as well. Many scientific activities that were seemingly devoid of politics in the past would raise the eyebrows of many of us today, including many scientists. I mentioned earlier how the Third Reich normalized the practice of experimenting on human beings as a *scientific* endeavor.[41] And in the United States, there was a time when the compulsory sterilization of "unfit" individuals was common, sanctioned and rationalized by scientists in the eugenics movement.[42] Many at the time would not have seen these activities as excessively political. Even today, certain types of research centered on maximizing human suffering and/or loss of life are equally normalized. These often fall under the euphemism of "defense research," like the development of weapons and military equipment, fueling armed conflict. In the United Kingdom, for example, institutions like the University of Bristol receive millions of pounds yearly in research funding from companies like Northrop Grumman and BAE Systems[43]—both of which specialize in creating machines for war. Some might say these are questions of *ethics*, not politics. But what is politics if not the contestation of ethical claims in the social arena? If weapons research is unethical, then why is it still allowed to continue? One could also say that eugenics research is unethical, but simply stating that claim does not explain why that type of research is no longer sanctioned in our universities. Political mobilization around scientific activities is the way ethical claims are given power and made to transform the social norms that govern which type of research is practiced, and which is not.

Beyond the avoidance of pain and harm, there are also many positive reasons why scientific research and political activism are frequent bedfellows. In his book *Impure Science*, Steve Epstein shows how activists played an important role in AIDS research in the 1980s and early 90s, shaping priorities and public views about the value of investigating the causes of this disease.[44] Moreover, Kelly Moore describes in her work *Disrupting Science* how many scientists took a political stance in the years after WWII, redefining their relationship to military and government institutions and making their science more transparent and democratic in the process.[45] And Albert Einstein— one of the scientists most revered in the popular imagination—was among the loudest voices for pacifism and against the proliferation of nuclear weapons.[46] His active engagement in the politics of his day made him a *better*, more thoughtful and passionate scientist than he would have been otherwise.

If we recognize, then, that scientific activities generally have some political valuation—for better or worse—then we open the door to other ways of doing science in the climate and ecological emergency. What could these involve? What would a break with science as usual look like today? Gianluca, Lou, and Adrien exemplify one way forward: changing their research practices in increased awareness of how those practices affect the world where they practice them. Their emphasis on being grounded and connected to others around them places them far away from the Mertonian research paradigm based on detachment that we saw in chapter 4, and closer to the Indigenous paradigm of knowledge as relation.

As several scientist-activists and critical theorists have argued, scientists can also perform research in support of transformational change, in collaboration with social and ecological movements.[47] For example, they can help activists gather evidence about the polluting companies a movement may be fighting, or they can help them develop new strategies to mobilize more people into action. They can also collect evidence, analyze data, and provide insights into what drives people to join movements, directly contributing to more effective activism. Through their science, they can contribute to a politics of change.

Science, then, is a political act: it always contributes to the societies in which it is practiced. As scientists become more conscious of their effects on the world, they can help better direct these practices toward positively transforming society. The slow-travel movement is one way through which they are more consciously contributing to these transformations, showing that not only are such changes possible, but they can also be highly rewarding. We can enrich our lives—as both academics and human beings—if we just take things slow, stop along the way to smell the roses, and maybe take a minute to talk to those caring for them.

CHAPTER 8

What Is It Like to Participate in Direct Action?

"[D]irect action is the defiant insistence on acting as if one is already free."

—David Graeber[1]

Scientist Rebellion's main modus operandi relies on breaking established laws or institutional norms. This is generally done openly, leveraging people's identities *as scientists and scholars*. In my experience, scientists are not prone to rule breaking, so they tend to be particularly averse to transgressing the law. At the same time, doing so can be very empowering, as feelings of hopelessness and anger are channeled into collective organizing for change. Indeed, an act of disobedience can generate a whole whirlwind of emotions: excitement, fear, anxiety, and even relief.

While in previous chapters I described different examples of scientist-activism, I have mostly focused on the look of SR actions, on how they were perceived from the outside. Here, I would like to delve into the inner workings behind their planning and execution. What happens to scientist-activists as an action is prepared, takes place, and eventually ends? How do they organize beforehand, or long after it has concluded? And how do they cope with the emotions that the action generates in these different stages?

Here, as I did in chapters 3 and 6 when talking about actions where I was involved, I will switch pronouns, from *they* to *we*. I will do this because I would like to include my own experiences and emotions as a scientist-activist—what *I* have learned about organizing and participating in direct action and what *I* have felt while doing so. So what follows comes more from a perspective of an insider participant than an outsider observant (with apologies to Merton).

FROM AN IDEA TO A PLAN

Let's start from the beginning: the organizing stage. Direct action requires quite a bit of organizing. We need to be clear and united in our goals and strategy. What are we trying to achieve? What is the action narrative? Who are we targeting? Why are we targeting them? Not only are there conceptual points to sort out, but there are also technical details: Are we blocking a street? If so, which one? How do we deal with stalled traffic, angry drivers, and the police? How can we make space for ambulances to pass through? All of this involves careful thinking, dialogue, and, above all, trust.

Activist communities have developed useful tools to foster all three of these values and make sure no voices are silenced. Scientist-activists rely on these tools when planning and executing an action. In Scientist Rebellion, the principle of "horizontalism" is highly cherished and respected. This principle was borrowed from Extinction Rebellion,[2] who in turn modeled it after the earlier Occupy and alter-globalization movements.[3] Actions don't have leaders; power in decision-making is distributed as equally as possible among the members of the group. Of course, some activists often take on larger planning responsibilities than others, but there are strong incentives for these weight bearers to rotate as much as possible. Everybody steps into an action as an equal partner and takes on as much risk as they are comfortable with. Ultimately, nobody should feel forced or coerced into a task.

To distribute roles and come up with an effective plan, we absolutely need secure communication channels. We try to meet in person when planning actions, and we turn off our cell phones when discussing critical details. We use end-to-end encrypted communication software. Particularly when we are discussing sensitive details, we avoid sharing compromising information in crowded channels, where police might be listening. We follow the principles of security culture: a set of guidelines that direct action groups have developed over years of fighting against police surveillance.[4]

Actions are often planned weeks, often months, in advance. And there's quite a lot of work that can happen in that time: designing leaflets, printing banners, building tripods and props, and even writing press releases with talking points for the media to use. All of this requires time and people power. It also requires logistical coordination: How do we get all the needed material? Who is in charge of each task? And how can we make sure everything is ready at the right time and the right place?

Most importantly, we need to get people on board. Acclimated as I am to academia, I used to think organizing an academic seminar series was complicated. Yet, mobilizing people for an action is on a whole different level. It can be easy to convince a professor to come give a guest lecture to a hall full of students: send an email to them, and you expect a positive reply about half the time. It is much harder to convince them to give the same lecture in the middle of a busy street, while activists are blocking it, and the police are arresting everyone around them! Thus, scientist-activism often involves sending a lot of invitation emails and getting a lot of negative-yet-polite replies: "I appreciate your eagerness to hear me talk, but I'd rather not do it in handcuffs." Over time, one builds a certain resistance to disappointment.

Other times, people surprise you. Scientists and activists you were not expecting to show up might suddenly appear days before an action, eager to lend a helping hand. And a helping hand in activism means the world. Even minor acts of kindness—like providing a house for us to meet and plan in, or a friendly hug before an action begins—can make a whole lot of difference in how a protest unfolds later on.

ALL HANDS ON DECK

After the planning stage is finished, the action day arrives. This is a time of emotional buildup: the minutes or hours right before an action can be nerve-racking: Will everything go as planned? How will the public respond? How aggressive will the police be? This is also the point where adrenaline starts to kick in, so it's wise to try to maintain calmness and to help others remain calm as well. I often remind myself to expect the unexpected. I have never been in an action where things went exactly as planned, yet that rarely meant that it wasn't successful, be that at gathering attention from the press, politicians, or members of the public.

And so, the action begins. We have trained for this moment, and each of us has a role to play. Some scientist-activists might be in charge of sitting on a road or occupying a building, while others might be taking photographs or video. Some might glue or lock themselves to a wall or to the pavement, to make it harder for the police to remove them. Others might mediate with the confused (and sometimes angry) members of the public, whose day is being disturbed. Some might talk to journalists on the scene or write social media posts with live updates about what is happening. And others might try to establish a dialogue with the incoming police officers, sometimes successfully, sometimes not so much.

Here again, trust is key. Scientist-activists organize in affinity groups—small teams of people (typically between 5 and 10) who know each other well, who can confide in each other, and who can work effectively under pressure. We need to know we can rely on one another to keep an eye on when there are arrests and to make decisions when things don't go as planned. Large actions usually involve several affinity groups with defined tasks: for example, when occupying a bridge, groups A and B might be in charge of blocking the two entry points from either side, while group C sets up hard-to-move objects (like lock-ons and tripods) in the middle, to buy time for the action to last for a long time. At times, delegates might be sent from all the groups to coordinate decision-making and relay information back to their respective groups.

My first international direct action taught me the impact that affinity groups can have when working in coordination, like groups of instruments producing a beautiful symphony. In the fall of 2022, over 80 scientist-activists from all over Europe had gathered in Germany for a new wave of rebellion. We knew we had the numbers to pull off something bigger than the Glasgow bridge blockade. A smaller team of scientists had for weeks been planning a series of disruptions we could carry out while we were there together. The first of these would be at the World Health Summit in Berlin: an international gathering of hundreds of researchers, medical practitioners, public health specialists, and politicians intended to shape global health policies. German Chancellor Olaf Scholz would give the inaugural speech. We planned to disrupt it.

Scholz had been recently criticized for his support of Germany's coal and car industries. Even as he styled himself as the "climate chancellor,"[5] his government was subsidizing fossil fuels to the tune of 21 billion euros a year.[6] Our action needed to draw attention to the connections between global health and climate breakdown, so we printed scientific papers from medical journals describing the increased risk of diseases brought about by the rise in CO_2 emissions. Then, we organized into affinity groups. I was in one of two groups tasked with plastering the WHO conference venue—a luxurious hotel in the center of the city—with the papers. Two other groups would block the entrances to the hotel, pasting their hands to the floor with superglue. A fifth group would try to enter the hotel and sound the fire alarm in the middle of Scholz's speech, while yet another group would stand on the street by the venue, holding banners for the media and the conference attendees to see.

Each group had a designated coordinator, who knew the exact time their team would have to carry out their activities. To minimize suspicion by hotel security until the very last moment, we arrived at the venue via separate means of transport (different buses and subway lines). And then, at the right time, we put on our lab coats and rushed to the hotel.

As I saw the other affinity groups, the adrenaline kicked in. I had never seen so many SR lab coats in one place. As I was painting the hotel's front glass with wheat-based glue, I could see fellow scientists blocking all the entrances, standing or sitting sternly in front of hotel security. Some had even managed to make it inside. Then, just as Scholz was taking the podium, the fire alarm blared. Irritated by his stolen spotlight, Scholz began by saying, "They are doing some protests about climate and things like that. I think the best way to improve our discussions is to just continue without listening to it."[7] *Don't listen to the climate alarm.* The "climate chancellor" could not have said it better. There was a host of journalists there already because of the speech, so we now had dozens of microphones at our disposal. We could tell the world about Scholz's fossil fuel policies and how they ran counter to all scientific recommendations, to all notions of a sustainable future. We could show the world the desperation we were feeling as scientists, as the canaries in Scholz's coal mine.[8]

During the whole action, there was an intense spirit of camaraderie: we did things together, and we took care of each other. We watched out for anybody finding themselves in uncomfortable situations. We had protocols in place for fast deliberation, in case the situation called for it. And we loudly supported those who were being arrested, to the tune of songs and mantras that we had learned in our respective countries. Nothing beats hearing the words "You are not alone!" shouted from afar as you enter a police van.

THE HUMBLING OF THE NERDS

Actions by academics have a particularly nerdy feel to them. The lab coat is a symbol that SR members are scientifically trained, that we can understand the science of climate breakdown. Yet many scientists who participate in these actions—myself included—rarely, if ever, have worn lab coats in their lives. I am a computational biologist. Most of my day is spent writing code on a computer. If I showed up to work wearing a lab coat, my colleagues would burst into laughter.

Yet, the lab coat has an empowering influence when worn in direct action. The media is drawn to arrests of people in full scientific paraphernalia, sometimes carrying books or journal articles as they are being handcuffed. It has an element of shock: a symbol of *the science* itself being silenced by *the state*. At the same time, the symbolism carries with it a certain tension, as it sets the scientist apart from the other protesters. This can have both positive and negative connotations, as expressed to me by Dr. Fabian Dablander, from SR Netherlands:

> In terms of visibility, I think it's quite powerful. When we join with our lab coats, the other protesters are very happy to see academics joining. For me, personally, at the beginning there was this tension where you put on a lab coat and there's some sense of "we're different form the other protesters"— a distinction I'm uneasy with. At the same time, it's signaling to the public that one of the most trusted groups in society is engaging in protest.

Dr. Nana-Maria Grüning—a molecular biologist and activist in SR Germany—told me how the lab coat affects not just others around her, but herself as well:

> I feel that people listen to me more when I'm with it than when I'm without it. There's also a difference in how I perceive myself. Putting it on reminds me of my education and gives me a sense of responsibility towards my own profession. I'm here, and I want to be taken seriously.

While scientific symbols are powerful, the particular modes of behavior we have come to value as scientists do not usually work well in activism. As we saw in chapter 5, capitalist academia trains us to behave in ways that do not readily translate to effective organizing or to generating consensus among diverse voices. We are encouraged to abide by the expectations of hierarchies, to compete and follow the rules, not collaborate to transgress them. Thus, we tend to rely on the humbling wisdom of experienced activists who are not scientists: people who have not necessarily dedicated their lives to studying textbooks, preparing experiments, or deriving equations but have obtained tons of knowledge from collective experiences. These "movement-based forms of knowledge" often prove more useful than academically derived knowledge.[9] This may be obvious to people

FIGURE 14. Dr. Nana-Maria Grüning (center) being interviewed by *Die Welt* and ServusTV, at the Letzte Generation Massenblockade Berlin, held on Straße des 17. Juni, Berlin, on October 28, 2023. Credit: Robbie Morrison.

working outside academia but perhaps not so much to those inside the ivory tower: actually doing something is much more conducive to learning than simply reading about it. Like nothing else, participating in activism helps people understand what works and what doesn't, what helps mobilize others and what drives them apart.[10]

The act of participating in direct action can also be humbling. By repeatedly mounting resistance and facing repression, scientist-activists come to understand just how entrenched the power structures they are fighting really are. Again, with Dablander:

> Changing systems requires collective action. Getting out there and trying to change policy or public perception—there is a level of insight that you get from that about the system, about how much perturbation it can absorb, that is more real, more profound than if you just stay in the ivory tower and write papers. If you're researching transformation but you're not involved in that transformation on a practical level, it's very difficult to get the level of insight you need. Activism can sober you up very quickly.

This way, scientists undergo "the political learning curve represented by social movements," which scholar-activist Laurence Cox argues can lead to forms of understanding that are more useful than academic theorizing.[11]

CAMPING IN A SPORTS CAR PAVILION

An action might be over before we even realize it: police might move fast to clear the area. The disruption at the World Health Summit lasted for a couple of hours, but actions can last for much longer than that. Just a few days later, Scientist Rebellion carried out a building occupation that stretched for 3 days. This time, the target was the German car industry, and the venue was one of the pavilions at the famous Autostadt, or Car City. Autostadt is an automobile expo located in Wolfsburg, an industrial town founded on Hitler's orders in 1938 as the city of the "Strength through Joy Car."[12] Originally meant to house the workers of Volkswagen factories building military equipment for the Wehrmacht and the Luftwaffe, the town was later repurposed as Volkswagen's trophy case. Now, the town is host to over a dozen giant pavilions where the latest car models are displayed for the world to see. It attracts over 2 million visitors per year.[13]

The plan for the occupation involved a team of scientists gluing themselves to the floor of the Porsche pavilion and pasting scientific papers about transport emissions along its walls. It was meant to expose the trophy case for what it really was: a testimony to the ease with which car companies can influence, even control, government policy, against scientific recommendations. SR and other activist groups were demanding a speed limit of 100 km/h on the famous Autobahn, the German highway system.[14] This would massively reduce the country's carbon output: 5.4 million tons of CO_2 would be saved every year, by official government estimates.[15] Additionally, it could make the number of car accidents plummet.[16]

At the time, Volkswagen was actively lobbying against this measure. They had channeled over 6.5 million euros to German politicians, and more than 3 million euros to EU representatives, in order to prevent the 100 km/h limit from becoming law.[17] In addition to obstructing the speed limit, lobbyists had pushed the government to increase the cost of public transport. The affordable 9-euro public transport ticket, established in response to energy price inflation during the Russian invasion of Ukraine, had just been discontinued, also under pressure from the car lobby.[18] During its short lifespan, the ticket had incentivized travelers to choose trains and buses, and it had prevented 1.8 million tons of CO_2 from being emitted.[19] SR wanted to draw attention to corporate lobbyism and to the importance of affordable public transport as a form of climate mitigation.

Once the scientists entered the Porsche pavilion, they were greeted by a confused team of security guards: why were these people in lab coats gluing themselves to the floor of the expo? When they caught up with the situation, Volkswagen chose to wait them out, in order to avoid a PR nightmare. So the activists stayed overnight.

Teresa Santos—at the time, a conservation biologist doing her PhD in Portugal—was one of the 16 occupiers. I reached out to her to get her take on the experience, and she told me what it was like to spend the night inside the pavilion:

> At first, the Autostadt staff wanted to seem as nice as possible. They knew journalists would pick up on the action if they called the police. Instead, they closed the pavilion and pretended to ignore us. After a few hours, when they realized we were not leaving, they tried to push us out. Activists that were right outside the pavilion faced violence from security guards, and one guard was even pulling their hair to get them to leave the premises. Inside the building, they turned off the heating and the lights. A patrolling guard would shine a lantern into our eyes and laugh, telling us to go sleep somewhere warmer.

That was the first night. The scientist-activists remained inside for yet another day and another night. Volkswagen only called the police on the third day. The extended length and conditions of the action generated both positive and negative feelings among the occupiers:

> It was a stressful situation, but there was lots of camaraderie among the activists and a strong sense of community. People were sleeping close together, making sure everyone was OK and feeling safe. We were sharing

jackets for those people feeling particularly cold. From a purely practical viewpoint, we couldn't afford to have people freaking out and leaving, as we needed the time for the media to show up. During the night, we were scared that the police would come in and arrest us by surprise, so sleeping was not easy. I had bruises on my knees because I was so cold my legs were bumping each other while trying to get some rest. We were also very hungry because we were not expecting to be there for so long.

An important aspect of direct action is to learn to improvise. As the hours passed, SR realized they had to find a way to deliver food into the occupied pavilion (which was surrounded by security guards). I initially did not take part in this action, but I had been in the SR action at the World Health Summit in Berlin a few days before, so I was near the area at the time. The activists who had been removed from Autostadt at the very beginning were looking for someone who had not been identified by security yet. I volunteered to try to sneak into the expo. I brought a bag full of snacks, bought a ticket, and made my best effort at acting like I was an avid car aficionado. Once inside, while pretending to take photos, I threw the bag of food into the closed pavilion before the guards realized I was not particularly interested in the latest Porsche model!

Teresa explained how they kept in touch with other activists outside the pavilion:

> After a day had passed, we couldn't charge our phones so as to keep in touch with others outside the building and tell them how things were going. The only place we could use electricity was the women's toilet, so the women in the group would surreptitiously go to the toilet and under the guards' noses charge everyone's phones.

Again, expect the unexpected.

NARRATIVE CONTROL

Even when an action seems to be over, work still remains to be done. Some people might be taken to a police station or detention facility. Being under arrest is stressful and tiring. We are temporarily deprived of our freedoms, and activists from minorities often have to withstand particularly abusive behavior from police officers. Those who are not arrested show gratitude and support by taking turns keeping watch outside the station, ready to welcome those arrested, with food, water, and open arms, after they are released.

In the case of Autostadt, the police arrested all the activists at dawn, using a back door to the expo so that no visitors would be able to see the arrests. One of the activists—a person who would rather remain anonymous—had been demanding medical attention for hours during the second night, due to extreme pain.[20] She was sent into a room and told to wait for paramedics to arrive. Unbeknownst to her, she told me, Autostadt staff had locked her inside the room:

> Eventually, two paramedics came into the room, but I didn't want them to treat me inside there. I just wanted to leave and go to the hospital. The Autostadt staff told me they would call a taxi for me, though I had said that was not necessary. When I tried to get up and leave, I realized the door was locked. The staff member that was in there with me pretended to be confused, and, soon after, four anti-riot police officers—with shields, sticks, and helmets—burst into the room I was in. At that point, the staff member quickly went behind them and locked the door again. Laughing, she pointed at the police and said, "This is the taxi you were waiting for." The police said I wouldn't leave until everyone else was also arrested. They eventually took me out in handcuffs to join the rest. We were kept outside, by a wall of the building, for a long time. I kept telling them I was in pain and needed to see a doctor. They instead took me to a police station for questioning and to take photos of me. Then, they sent me to a second station. While they were questioning me again, I remember falling to the floor in pain. I started crying because I couldn't resist any more. At that point, many hours had passed since I had asked for medical help. I think one of the female officers saw me and convinced the rest I needed treatment. They finally took me to the hospital in another cop car.

In a statement posted on their website, Volkswagen claimed they had treated activists humanely, with "care for their health."[21] This is where narrative control is particularly important. Journalists are hungry for quotes, and they are especially in need of a story to tell. How that story is told (and who gets to tell it) can make a big difference in how the public perceives what has actually taken place. If there had been nobody on standby, prepared to engage with the media, Volkswagen's narrative might have gone unchallenged. Journalists might have defaulted to getting quotes from security guards, corporate executives, or the police. To counter misleading narratives, designated spokespeople in activist groups are trained and prepared to talk to journalists, to explain the reasons for the action and transmit the demands to the public.

A social media team is often in charge of connecting with the public *live*, delivering photos or testimonies from the scientists who are risking arrest. For example, during the pavilion occupation, the Autostadt CEO came to talk to the scientist-activists on the first night, telling them that "the course Volkswagen is taking is one of sustainability." The irony was apparently lost on him that among the occupiers was an actual environmental scientist, Nate Rugh, who challenged him on this lie. Thanks to the social media team, it was not lost on the hundreds of people who saw them talking on their Twitter, YouTube, and Facebook feeds. Live testimonies of the occupiers helped to put a human face on the occupation. Dr. Marta Matos, another scientist at the pavilion, delivered a speech while glued to the floor:

> I'm here because, as a scientist, I know that Volkswagen's greenwashing and lobbyism is killing people. The carbon emissions they are responsible for are causing deadly floods, wildfires, heat waves and droughts, like what we've seen in Europe this summer. And it's much worse in the Global South. Soon, whole regions will probably be uninhabitable, and that's a crime.[22]

Thanks to extensive back-office work by SR members who were staying on the outside, the Autostadt event garnered attention from traditional media as well: a large number of local and international news outlets reported on the event.[23] Dr. Nana-Maria Grüning was one of the people talking to journalists while the activists remained inside the pavilion:

> The biggest challenge was to find and deliver good talking points for the journalists to use and to make sure the pressure was sustained for long enough so that enough outlets would pick up on the event.

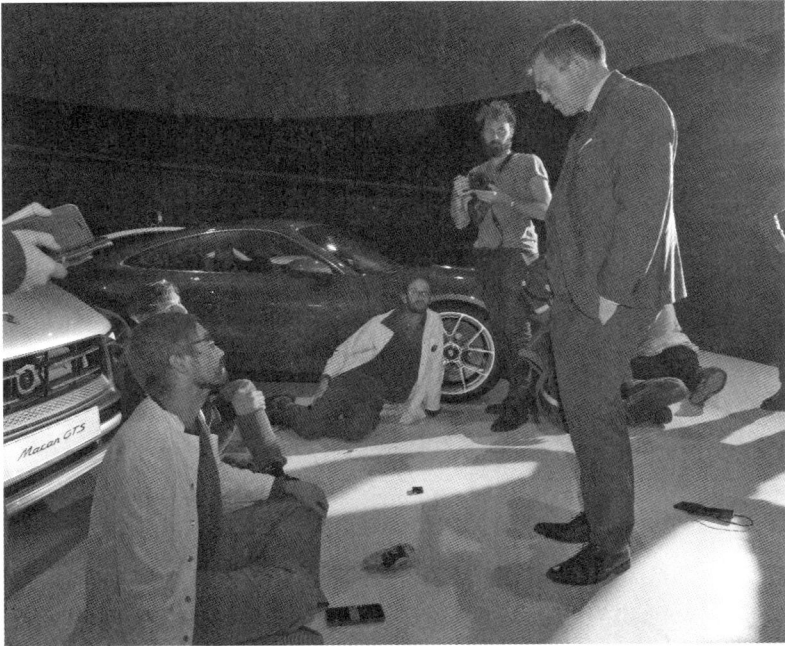

FIGURE 15. Scientist Rebellion members in conversation with the Autostadt CEO inside the Porsche pavilion. Credit: Isaac Peral.

DISRUPTION AS CREATION

Interview requests kept arriving long after the Autostadt action was finished—from journalists as far away as Brazil, Turkey, Uruguay, and South Korea.[24] Eventually, as things settled down, we realized it was time for a debriefing session—a must after every action. Debriefing sessions might happen hours, days, or weeks after, but we don't consider an action truly finished until we can meet again and talk about what happened. We collectively reflect on what we liked and disliked, what worked and didn't work, and what we can do to improve moving forward. Ultimately, every action is a learning experience.

These sessions also serve as an emotional outlet: we relive together the event, sharing how we felt throughout. To me personally, there is often a feeling of elation that comes with direct action, a feeling that I was—at least for a brief moment—doing something with a deep meaning and of great consequence. There is a recognition that my peers and I managed to break the silence of convention, the veil of normality, and highlight that, no, we don't live in normal times, and we shouldn't act like we do.

Teresa recounted her own feelings a year after her arrest in Wolfsburg:

> In some ways, it has affected me more as a person than as an activist. It's gonna sound silly but before the Autostadt occupation, I used to be terrified of public speaking. Now I think to myself, "If you managed to survive that, you can surely survive a PhD defense." It's weird, but direct action to me is a good cure against anxiety! It helps me to put into perspective what's important and what's not—and to build connections and friendships.

Direct action is thus a profoundly transformative experience for those involved. On the one hand, it is a form of "prefigurative politics": a way for activists to tell the world that societal change *is* possible, by actively creating that change in the process.[25] On the other hand, scientist-activists change *themselves* while engaging in disruption, in ways that affect them well after an action has ended.[26]

In the case of Germany, laws cracking down on transport emissions are entirely possible—but corporate interests remain in the way. The actions by SR exposed those interests and generated a societal discussion about the influence of the car industry in political decision-making.[27] Disruption is, ultimately, a form of creation: an embodiment of the world we aspire to reach and, simultaneously, a denunciation of the barriers preventing us from getting there.

CHAPTER 9

Reflections Inside (and about) Cages

"There's no justice in this system
A bigger cage is still a prison
Well I hope that somewhere someone breaks free tonight."

—Cistem Failure[1]

Just a week after Autostadt, Scientist Rebellion was ready for round 2. Their overnight occupation had caused a media storm, and they had the human power to escalate even further. They remained fixed on their demands against German transport emissions, but they knew that occupying the same car pavilion twice would be difficult. They needed to move locations. This time, they chose to travel south and target another staple of the car industry: the BMW Museum in Munich.

On the morning of October 29, 2022, 15 scientists entered the museum and glued themselves to a luxury car inside, after throwing removable paint on its chassis.[2] Many of them had already been arrested earlier in the Autostadt car expo in Wolfsburg. One of them, Lorenzo Masini—a plant biotechnologist from Italy—told me:

> After finishing my MSc degree, I took some time to think about what to do next. I had a chance to start a PhD, but I decided to instead put my energy into activism. I moved to Germany for a month to help organize the Scientist Rebellion campaign there, which culminated with that last action in Munich. We targeted the BMW Museum because it's a symbol for the fossil-fueled car industry.

FIGURE 16. Action by Scientist Rebellion at the BMW Museum in Munich. Credit: Isaac Peral.

This time, all the scientists, including Lorenzo, were arrested immediately. Back in Wolfsburg, they had been released from the station after a few hours. In Munich, they were imprisoned for much longer than that:

> We were kept in a police station for two days. I was in a small room without a clock and without windows. Just white walls, a mattress, and a toilet. We only saw the people giving us the food through a little hole. Then, at 2 a.m. of the first night, I was presented in front of a judge and sentenced to stay for 5 more days in another facility.

A media campaign immediately followed, this time with testimonies of Lorenzo and the other scientist-activists—biologists, ecologists, physicists—who were imprisoned. Scientist-activists held solidarity protests in German embassies across multiple countries, from Norway to Colombia, holding photos of the arrested scientists. Quotes from the arrested scientists made it to newspapers around the world.[3]

SR was trying to underscore the extremes to which the German government would go to protect its automobile industry.

BEHIND BARS

As we saw in chapter 2, scientist-activists are challenging the cultural hegemony of fossil capital. In doing so, they face that other force through which power perpetuates itself: not culture, but physical coercion. They come to learn that police, private security, and the prison system as it is currently built are not there to protect them, or the living world, but to protect capital, to prevent any threats to the status quo—the ways things are today. *Never mind these scientists and their warnings; get them off here so I can keep selling cars.* This is also part of what sociologist Laurence Cox calls the "political learning curve" of social movements. Scientists often come from privileged backgrounds, and so, unlike historically marginalized groups, many of them lack an initial distrust of police. During direct action and its subsequent repression, all of this changes.

In this chapter, I would like to reflect on the experiences of scientist-activists in jail. What goes through their heads once they're inside? How do they cope with the seclusion? And what do their fellow activists do while they're in prison? I will also use this chapter to evince the many connections between jails and climate breakdown. Just as in chapter 8, I find myself alternating between the first and the third person: I have also been arrested, so the reflections below will swing between being about "them" (as an observer of those scientist-activists choosing to risk arrest) and "us" (as one of the many scientist-activists who have been arrested).

Scientist-activists who carry out civil disobedience do it in full knowledge that police will act to detain them. Thus, you will be hard-pressed to find SR members feeling repentance or regret during confinement. As Lorenzo explains, time is better spent in other matters:

> In that situation, you can't really do much more than sleep, stretch, and sometimes sing. In the jail in Munich, we sang many songs, and the songs

carried through the walls. We formed a sort of chorus with the other imprisoned researchers. Later on, they gave me pen and paper, so I started writing down plans for future actions to do with my local SR group in Italy. I was also figuring out ways to mobilize more scientists after I came out of jail.

In prison, the mind wanders, sometimes in unexpected ways. For Lorenzo, perceptions of time changed dramatically:

> You have a lot less inputs inside. No smartphone, no computer, no watch. Not even a clock in the room. It's not something we are used to these days, without thousands of inputs coming at us all the time. It's a bit weird, but after a while I was feeling more relaxed. I took time to think about my own life: stuff that I wanted to do after I got out, and people I wanted to meet. In some sense, time had a different "speed." With so little input, your brain is not good at recording the fact that time is passing.

Scientist-activists still have worries. Sitting inside a cell makes one think about those outside: concerned family members and friends. Did they see the action on TV? Do they know that we are OK?

> While inside, I was often thinking about the people that I love. There was one beautiful day when we received all these letters, because of a solidarity campaign SR was carrying out. Each one of us received their own letters. I got one from my sister, one from a friend, and then a whole lot from other activists. There were also random people in the world writing to us beautiful things, including poetry! Some just said "Thank you for what you're doing. I have kids, and it's good there are people doing this for us." It was an intense and emotional moment for me.

Thus, like a modern-day Gramsci, Lorenzo stayed connected with the outside world through these letters, while keeping himself busy ideating ways to mobilize more scientists into action. His experience is but one example of what happens to activists behind bars. Experiences in jail are of course different depending on the circumstances and locations of the arrests. Sometimes, these can involve large amounts of violence and abuse. In countries in the Global North, groups like Extinction Rebellion, Just Stop Oil, and Last Generation have amassed large teams of volunteer lawyers and legal experts, who provide advice about potential risks to those breaking the law.

This way, activists like Lorenzo know what they might be getting into, way before they are taken by police. This, in turn, reduces some of the anxiety that inevitably comes with the prospect of confinement.

Yet, there is no guarantee that police will act by the book and follow established procedures. In particular, marginalized minorities and Indigenous groups are often subject to police brutality over their environmental activism, at much higher rates than white activists experience. And in some countries, arrest can involve the risk of death.[4] Jordan Andrés Cruz is a tropical ecologist from SR Ecuador, for whom the idea of prison has a very different connotation from what Lorenzo experienced:

> In Ecuador, going to prison is essentially a death sentence. One often hears about prison massacres, due to drug dealer bands fighting for territorial control within the facilities. And people here can remain locked up without a sentence for years.

In other countries, it is the risk of contracting deadly diseases that governments utilize as a deterrent against protests, as Shami—a scientist-activist from SR Uganda—explains:

> Here, staying in prison for just a few days is enough punishment. The toilets, the conditions, make people scared; cholera is widespread, so they don't expect you to campaign again after an arrest. That way, the government handles resistance through repression.

The heightened risks might also mean that scientist protests in certain countries have to be carefully crafted. Jordan continues:

> When the police arrive, we try to explain to them that we are academics from universities and NGOs and that we will leave within a specific period of time. We try to calm them down and avoid arrests, and our credentials tend to help with this, reducing police aggression towards us.

Jordan points out there is a "credential privilege," which is also something scientists in many countries have noticed, even in the Global North. The lab coat tends to prevent police from handling scientists the same way they do other protesters. I have often been told by nonscientist activists how surprised they were at the way police were treating me and other academics. This is a privilege that marginalized communities do not have. As Jordan tells it, "this makes our responsibility particularly important: as scientists, we have the obligation to be louder and speak for those who cannot."

THE CAGES OF MODERNITY

Partly due to my credentials and to the fact that I live in Northern Europe, I have been treated fairly leniently when under arrest. During those times, I have had the chance to wonder about the very concept of cages. To me, one of the biggest indictments of our world is that it *is* a world of cages. Not just prisons, but forced-labor camps, reeducation facilities, migrant detention centers, and psychiatric institutions. The words to describe cages are rich and diverse— testimony to the efforts governments will go to in order to pretend they are not, in fact, putting people inside them.

Cages are also the way most of our food is sourced: factory farms, where animals spend their lives in tiny spaces, are pumped for milk, are killed for meat, or are forced to raise the next generation of animal prisoners. The cages made for plants—monoculture zoning and fencing—prevent species from interacting with each other, depleting the land of nutrients and promoting desertification. Ecological breakdown is rooted in these cages. Agricultural enclosures were the original way by which capital secured profit, through the commodification of nutrients and minerals contained within them. Starting in the seventeenth and eighteenth centuries, commons that had been shared and cared for collectively were fenced off, often by violently forcing farmers off the land.[5] Contrasting this to the common myth of the "tragedy of the commons," the writer Amitav Ghosh describes it as a "tragedy of enclosures": the imposition of private property regimes on vast areas of the world, first in Europe and then everywhere else.[6] It was these new cage regimes that eventually catalyzed our dependence on fossil fuels. In preindustrial Britain, lords began to appropriate areas wherever they suspected precious minerals (including coal) were buried. These lands turned into the mines that would feed the steam engines of the nineteenth century.[7] Cages were thus instrumental to the birth of the fossil economy.

Spatial enclosures have not just reshaped the countryside, but also come to define urban life. Over the twentieth century, cities around the world were progressively mutilated by asphalt, through the construction of avenues, parking lots, and highways. In the name of car-centric mobility, the concept of the "pedestrian" was invented. Those without cars became second-rate citizens, always in deference to the speeding driver and at risk of "jaywalking" if not moving at the right time and place.[8] The hegemony of the car was thus (literally) cemented in the public imagination.[9] Now, it is hard to think of urban life without the sounds, sights, and smells of motor vehicles. Cities became cages for those who did not own cars.[10]

In many ways, cages are the reason that climate activists are risking being put into cages: mines, highways, industrial farming—all forms of enclosures, designed to extract profit. But we need not get so poetic: environmental justice demands often intersect with the demands of movements against the most familiar kinds of cages, movements that seek to abolish prisons and policing, in favor of other forms of public safety. A clear example of this is the Cop City standoff in the US state of Georgia. There, a massive police training center is under construction, in an effort to increase police militarization in a predominantly Black neighborhood, just 2 miles southeast of Atlanta. The center's planned site is in the middle of a lush forest ecosystem, designated by the city government as "one of Atlanta's lungs."[11] Thus, construction of the police facility has involved the felling of hundreds of trees. Here, ecological and abolition activists have found common ground and joined forces, through a decentralized movement called Stop Cop City. They have mounted a formidable defense since 2021, building barricades and tree camps inside the forest and sabotaging construction equipment.[12] Many of the protesters have been charged with domestic terrorism,[13] and one of them—Manuel Terán, known also as Tortuguita—was killed by police during a raid against the encampments.[14] Scientist-activists like Rose Abramoff (whom we met in chapter 4) have joined local elders and other activists opposing Cop City, mounting protests against the company carrying out the facility's construction.[15]

Cop City exemplifies how the drive to police militarization and incarceration tends to clash with demands for ecological justice. But ecological justice is not just about struggles against physical enclosures. Today, activists also face temporal cages as well. These include modern concepts like the strictly timed working week and modern values like the economic glorification of salaried work.[16] Though fairly recent, these ways of conceptualizing our daily lives have become normalized—they are a key component of the fossil economy. As ecological economist Erik Gomez-Baggethun explains, the idea of selling one's own *time*, of being strictly *on the clock*, was another modern invention: a way for capitalists to exploit the labor of those who had been denied access to common land.[17] Preindustrial societies were not organized around work as paid labor.[18] And throughout history, most free people have tended to sell their labor (or had it appropriated) in the form of physical output: number of fruits collected, ounces of milk produced, etc. In contrast, in the modern world, the selling of one's entire time ("I work a 40-hour job") is largely taken for granted. And when almost every activity is appointed, scheduled, or reserved—even beyond nominal working hours—free time becomes a scarce resource. This affects psychological well-being, while restricting our potential for societal engagement. As Ajay Singh Chaudhary stresses, "extending and accelerating productive time as far as possible, well into supposed non-working hours, not only generates profitability but in the process specifically destroys the time for politics."[19]

Scientists are also becoming prisoners of their own schedules. Managerialism and hyperaccountability are choking the potential to do anything meaningful in the academy: ever larger chunks of working time are devoted to writing reports to supervisors or managers, stating that one is doing what one is supposed to be doing.[20] These self-reporting tasks, in turn, stimulate the creation of new tasks to report on the reporting acts themselves—what David Graeber calls "bullshit tasks."[21] Thus, academics fall down ever deeper holes of (self-)surveillance and paperwork.[22] This is one of the most frequent reasons given by scientists for not engaging with causes that matter to them, including climate activism.[23] Along these lines, Max Weber talked about the "iron cages of rationality": the mental and bureaucratic structures that trap people in social arrangements defined by efficiency and mechanization.[24] Scientist-activists are trying to deconstruct these cages as well—for Scientist Rebellion, mobilization involves encouraging colleagues to challenge and break out of temporal prisons, for example, by lobbying for activism to be a valid form of academic work. A recent article by scientist-activists stresses the urgent need for universities to "explicitly recognize engagement with advocacy as part of the work mandate of their academic staff" and to support their staff so that they can "redirect their work on campus to assist the social movements addressing the climate and ecological emergency."[25]

Scientist-activists are also struggling against prisons of the mind: the ideologies that prevent us from thinking another world is possible. We saw in chapter 1 how IPCC models, for example, delimit what scientists and policymakers consider plausible and how these models are based on assumptions that are themselves rarely questioned: the continued existence of fossil capital or the reliance on technological solutions yet to be invented, to name a few. And there are even more pernicious mental prisons. We explored these in chapter 4: the "rational" constructions used to keep academics disengaged, passive, *neutral*; the arguments given to keep them locked up in the ivory tower.

To summarize, then, the cages of modernity—the ways in which societal engagement is contained, constricted, and suppressed—can be literal prisons, but they can also be prisons of our mobility, our time, and even our minds. All of these inhibit the work of climate and ecological activists, no less that of scientist-activists. Still, there is one more type of prison that we have yet to talk about—perhaps the cruelest of them all: the type we draw on maps.

A BIGGER CAGE IS STILL A PRISON

Borders are everywhere in the modern world, and they are especially omnipresent in political discourse. They are at the center of debates about war, migration, trade, and diplomacy. They frame these debates, constantly shaping the way people talk about their fellow humans. Borders allow the power core to divide what's "ours" from what's foreign, separating us from the stranger, the *Other*. Thanks to borders, those within them can devalue those outside them, can consider them less worthy of compassion, less like fellow citizens and more like invaders at the gates: a constantly looming threat. Thus, in the name of security and stability, borders justify violence and *indifference* to violence.

In the early twenty-first century, borders are becoming harder and harder for people to cross, even as capital becomes more and more fluid. The dominant class, the "jet-setters," can travel the world in a matter of hours, or they can make their money do the travel for them. Yet, the wage labor that capital exploits remains mostly fixed in space.[26] The average human has perhaps never had so many legal barriers preventing them from just roaming free.[27] Despite evidence that relaxing borders could increase the world's ability to cope with climate breakdown,[28] the highest CO_2-emitting countries are instead strengthening them. More than twice as much is spent on arming borders than on climate finance: money that could be going toward adaptation is instead invested in building weapons and walls.[29] Refugees from climate disasters and wars over resources face the violence of militarized frontiers; looking for nothing more than a chance to live, they are denied even that.[30] Climate breakdown is thus tied to control over human bodies: what they can do and where they can go. As the power core lays waste to vast areas of the world, it prevents those directly suffering the consequences from seeking refuge elsewhere.

The notion that a border is also a cage is nowhere more evident than in territories under colonial occupation: entire countries effectively run as prisons. Theoretical physicist Mike Lynch-White—the cofounder of Scientist Rebellion, whom we met in chapter 1—is one of an increasing number of activists working to bring those cages down. On May 27, 2023, Mike was sentenced to 27 months of prison—longer than any scientist-activist had been jailed before him. The sentence was given not for his work with Scientist Rebellion, but for another act of civil disobedience he committed, just months after the paper-pasting event at the Royal Society. That time, Mike was not wearing a lab coat, but a keffiyeh scarf.

On June 10, 2021, together with two other activists from the group Palestine Action, Mike occupied a factory in the industrial town of Runcorn, near Liverpool. The factory is owned by the multinational APPH and supplies parts to Elbit Systems: Israel's largest weapons manufacturer. Dressed in flashy red boiler suits, Mike and the other activists managed to climb atop the roof in the early hours of the morning. Once on top, they began breaking instruments and tools used for building drones. The activists remained on top of the factory for 36 hours, until removed by police. They were initially charged with £4 million worth of damage, though the figure was later revised down to £198,000.

Mike wanted to sabotage the Israeli weapons industry, but he also wanted to bring attention to its victims. The occupation of Palestine is one of the deadliest cages ever built. It was consolidated during what is called the Nakba—the forceful eviction of over 80 percent of the Palestinian population. Dating to 1948, when the nascent State of Israel moved to take over most of the Mandate for Palestine, the Nakba involved the killing of thousands of Palestinians.[31] Since then, the occupation forces have maintained their grip on the region through the displacement and killing of the remaining inhabitants, while treating Palestinians within occupied territories as second-class citizens.[32] Just in the two decades between 2000 and 2020, over 10,000 Palestinians died at the hands of the Israeli military forces or were murdered by illegal civilian settlers, who are in turn backed by the military.[33]

Today, the two biggest remaining bastions of the Palestinian State—Gaza and the West Bank—are open-air prisons of extraordinary scale.

Since 2007, access to water, food, and medicine in Gaza is under strict control.[34] No supply vehicle can enter without express Israeli authorization. Gazans are not even allowed to collect rainwater or build wells on their own land.[35] Meanwhile, West Bank villagers are regularly displaced or killed; their possessions and homes are destroyed to make way for new Israeli settlements.[36] Palestinians are seen by both the State of Israel and Israeli academia as lesser humans at best and "animals" at worst,[37] living in what Amnesty International calls a state of apartheid.[38] Mike puts it this way:

> Before the action in Runcorn, I had been to the West Bank, and I had experienced firsthand how kind the people there are. In fact, I don't think I would have become an activist if I hadn't been to Palestine. The amount of evil going on there is just unimaginable. I was then driven to let the truth be known, both about Palestine and the climate crisis. I am convinced that the world is better when people know and act on the truth.

Yet, he recognizes Palestine is unlike anything he could have experienced while in jail:

> Being born into a open-air prison, for the crime of being Palestinian, is something very different from what I went through. I knew I would get out at some point. What Palestinians are going through is vastly worse. I was safe in my prison, relatively speaking. There weren't rockets being dropped on me every day. It made my time in jail feel like less of a high price to pay.

Mike was imprisoned just months before one of the deadliest phases of the occupation. Triggered by an attack by Hamas on military bases and settlements around the Gaza border wall, the Israeli military launched a massive bombing campaign in early October 2023.[39] Over the course of the next months, Israeli forces advanced into the Gaza Strip, destroying more than 70 percent of its civilian infrastructure[40] and killing over 46,000 people at the time of writing[41]—about a third of them children. A study in mid-June 2024 estimated that, due to underreporting, the death toll may actually exceed 186,000.[42] Additionally, over 4,000 people (mostly children) have suffered amputations as a consequence of the invasion; it is estimated that around 10 children have lost an arm or a leg every day since the campaign started.[43] An almost complete food blockade around the Gaza Strip has led to over 1 million people facing what the UN calls "catastrophic levels of food insecurity"[44]—Gazans have been forced to starve even as they seek shelter from the bombings. Given the magnitude and sheer ruthlessness of the devastation, the UN special rapporteur of the occupied territories has called the Israeli campaign a genocide,[45] as has the International Court of Justice[46] and 100 civil society organizations.[47] The campaign has also destroyed the cultural, scientific, and medical support systems of the region. Through a combination of carpet bombings, controlled demolitions, and bulldozers, Israeli soldiers have destroyed all 36 hospitals that once stood in the Gaza Strip,[48] as well as every university and institution of higher learning.[49] They have also used sniper attacks to target healthcare workers,[50] scientists, and scholars.[51]

The genocide in Gaza has deep ties to fossil capital. On October 30, 2023, just weeks after the invasion had begun, Israel granted 12 licenses to six companies—including fossil fuel giants BP and Eni—to explore for gas off the Palestine coast.[52] Later that year, on November 20, the US energy secretary was sending special advisors to Israel.[53] The area's fossil reserves were valued at $524 billion. A large part of them lies in what the UN recognizes as territory of the State of Palestine. Yet, as they do with rain and groundwater, Israel also prohibits Palestinians from obtaining oil and gas from their own land.[54]

In addition to the lives of those directly killed, huge long-term climate costs have resulted from the campaign in Palestine. During the first 60 days of the invasion, the US and Israeli military forces emitted at least 280,000 tons of CO_2 into the atmosphere, exceeding the annual emissions of 20 small countries. This was mostly due to bombing raids by Israeli F-16 jet fighters, as well as US flights delivering ammunition. The actual emissions may end up being much higher (perhaps 5 to 8 times higher) once the full supply chain of the materials is included in the calculations.[55] On top of that, these numbers fail to account for the uprooting of thousands of trees and the destruction of agricultural soils through seawater infiltration,[56] as well as the carbon emissions from the eventual reconstruction of Gaza.[57]

THE BRIDE IN RED

Beyond its carbon "costs," the genocide in Gaza has involved unimaginable atrocities. It has been the first live-streamed genocide in history. Through YouTube, Instagram, and TikTok, the world got to see example after example of the carnage: parents using plastic bags to carry the body parts of their dead children; orphaned babies crying over the rubble of their former homes; dead bodies of toddlers still holding on to their last candy bars, minutes after a bomb removed all the oxygen from their lungs.

During the last months of his imprisonment, Mike could do nothing but watch these events from the TV set inside his cell. He felt powerless in the face of the destruction:

> They say that when you're put in prison, you're really taken out of society. You're genuinely not part of society. You feel like you're in a different universe. It's so all-consuming. All you know is prison. All you deal with is prisoners, and prison officers, and prison admin. When the genocide in Palestine picked up, I felt very removed from it, almost numb. I felt like I kinda knew that there was nothing I could do, beyond writing some statements so that somebody in the outside world would read them.

Thankfully, there was someone in the outside world making sure Mike's plight—and the much harder plight of those he was fighting for—did not go unnoticed. Like Mike, his partner Elena is a tireless scientist-activist, and she was moving heaven and earth to amplify his message, to bring attention to the injustices in Palestine for which he had risked so much jail time. As she tells it, "during all the time that Mike was in prison, we were trying to get him to the news. I kept thinking, *We have to make the news; we have to make some noise.*"

Elena was originally trained as an astrophysicist. Yet, both climate activism and Mike's imprisonment had put her academic life on hold. Studying the movement of stars and planets no longer had the same attraction it originally did—her own planet demanded her full attention. Elena's personal life was also in stasis: she had planned to get married to Mike, but his sentencing threw those plans overboard. For now, the ceremony would have to wait at least until his release.

At the same time, Elena was also fighting against state repression in her own home country of Spain. She had been one of the scientists who had thrown beetroot juice at the stairs and walls of the Madrid Congress, representing the blood of all those dying due to climate breakdown, during one of SR's coordinated days of global actions.[58] Even though the juice had been removed the next day, she and her fellow scientists were being charged with "damage to a historical building." If found guilty, she was looking at 1 year and 9 months in prison. This was part of a broader pattern of criminalization of climate activists. At the time, the Spanish attorney general had announced that civil disobedience groups would be classified as "terrorists,"[59] and the police had been found planting covert operatives inside these groups, to extract information from them.[60] Even some scientists who Elena knew personally had been targeted by the secret police.

State repression—in Spain, in the United Kingdom, and in occupied Palestine—was a common thread running through Elena's life. It was bearing heavily on her. Then something clicked. She realized this was a thread she could pull to gain public attention about repression in all three countries, all at the same time. She told me:

> Thinking back on my intention to marry Mike, I came up with this idea of an action that involved wearing a wedding dress. It could be an iconic image. Newspapers wouldn't be able to resist that. I was inspired by the *Hunger Games* series of dystopian novels, where young people also face heavy state repression. Like the main characters do in the novels, I wanted to use our relationship as a weapon.

Elena went all in with the bridal symbolism. She staged a theatrical civil disobedience action at the British Embassy in Madrid. Wearing the white dress she was hoping to use in her actual wedding, Elena and two fellow climate activists threw red paint at the embassy and pasted photos of Mike on its walls, demanding an end to state repression against activists. Her hands splashed with the color of blood, she was then handcuffed and arrested. The "bride in red" made

FIGURE 17. Elena being arrested after her action at the British Embassy in Madrid, June 16, 2023. Credit: Alberto Astudillo.

headlines across Spanish media, and it got people to talk about repression of climate activists, but also about Mike's imprisonment *and* the complicity of the British government in the plight of the Palestinians—all three in one go.[61]

THE SILENCE OF OUR COLLEAGUES

Today, Mike and Elena's life remains on hold. Mike has been released from prison, but he is due for a second trial for another climate action. Meanwhile, Elena awaits her own trial in Spain. They are among the bravest climate activists I know, yet it is easy to forget that they were once full-time scientists. When I met them for an interview, I couldn't help but feel a sense of shame for academia. Most of their colleagues have remained silent during their struggles against police repression. None of the universities that once valued them as knowledge-producers have made statements about their activism.

It is this silence that stands out. To a lesser degree, it is a silence scientist-activists face when we go *back* to the lab and try to speak about what happens to us during direct action. I will never forget that on the day after I was first arrested, while describing to my cow-orkers what it was like to be inside a cell, I soon realized they wanted to change the subject, to move on to weekend plans or the weather, anything to avoid talking about cages. Like the silence of the climate emergency—the "socially organized denial," the "double reality" of academia—the silence surrounding activist repression is equally dangerous, if not more so. This is why, like Elena with Mike and the SR campaign after the BMW action, scientist-activists make such an effort to bring attention to those behind bars; they want to break the silence of repression.[62]

This is yet another connection, perhaps a crucial one, linking climate breakdown with the world of cages, because cages, like oil pipelines, coal mines, and highways, are not just built with steel and cement. They are also built with silence. The silence of our colleagues and our friends. The silence of those who are not yet rising up in resistance. A silence that speaks louder than words.

CHAPTER 10

The Privilege of *Rather*

"Colonialism is not in the past, it is in the present and in the future. We are altered, reconfigured, expendable, but are not passive agents in this, despite constrained circumstances. We live, resist, rebuild, rejoice, and refuse. Yet, we also feel sorrow for a past that never got to be, a present that is incomplete, unknowables that haunt and pique. Our memories are reshaped and respond to local and global forces, we are all different but we share some common histories. These are the fertile grounds where colonial and imperial wounds and resultant rage, grief, and desire are not minimized but recognized as part of the driving forces of resurgence and liberation."

—Farhana Sultana[1]

"I'd rather be doing research in my lab, instead of here." I said these words once to a camera as I was being arrested. Just a few days before the Autostadt occupation, I was blocking the road in front of the German transport ministry in Berlin, together with 60 other scientist-activists. Our demands were the same as in Wolfsburg and in Munich: a speed limit on German highways and the return of the 9-euro public transport ticket.

It's not uncommon to talk about our personal lives while participating in activism. It is a way to humanize ourselves in the eyes of the public. We are not *just* activists. We are scientists, doctors, workers, parents, grandparents. We have lives outside of activism, and those lives motivate us; they are part of the reason we are stepping into the streets.

But I want to do some self-criticism here—not of the fact that I have a job as a scientist or that I, and other scientists in the Global North, am now engaging in direct action after years of appeasement of governments and corporations, but of the fact that I am able to be both an activist *and* a scientist, the fact that I can fulfill one type of role at a given time, *rather* than the other. This is a privilege many activists do not have, especially in the Global South.

For many peoples in Africa, Latin America, and South Asia, climate collapse is already here. While the year 2023 saw a record number of extreme weather events all over the world, Africa was by far the hardest-hit continent. The deadliest storm of the year was Storm Daniel in Libya, which killed at least 4,361 people and left $7 billion in damages—the costliest weather disaster in African history. It was made up to 50 times more likely and 50 percent more intense by climate change. That same year, Africa was also witness to Tropical Cyclone Freddy, its deadliest tropical cyclone on record, which ravaged the southern part of the continent, killing at least 1,400 people.[2] But dealing with these increasingly frequent catastrophes is not easy. Centuries of colonial and neocolonial exploitation have decreased the ability of communities to cope with extreme weather events.[3] Climate injustice thus projects from the past into the present and future: those with the least responsibility are the ones who are made to suffer most.[4]

The Global South is also where the biggest frontiers of capitalist extraction are located, the sites of massive projects set up to siphon resources from the land and the water. Not just fossil fuels, but the minerals to build solar panels, electric cars, and smart phones; the means for Europe and North America to achieve their "green transitions" while leaving the rest of the world impoverished. In her book *This Changes Everything*, journalist Naomi Klein combines the many groups resisting extractivism under the banner of an emerging "Blockadia" movement[5]—the global front lines against the advance of capital. For the environmental defenders in these front lines, activism is not a choice; it's a necessity. When a multinational decides to build a pipeline across your house, or a mining company evicts you from the land of your ancestors, there is no option but to organize and fight back. Thus, for many in these front lines, there is no *rather*.

What roles do scientist-activists play in these front lines of extraction? How are they working with other sectors of society that are affected by the expansion of Global North capital? And how are they coordinating with other movements and campaigns that also seek to stop this expansion?

STANDING UP TO BIG OIL

Esteban Servat is an example of a scientist-activist for whom *rather* no longer exists. Like me, he is an Argentine biologist, but he no longer works on research in that field. Esteban became famous for leaking secret government information about oil fracking operations in the Mendoza province of Argentina. In 2018, a report he released to the public showed that the government knew that planned oil extraction in the region would pollute the drinking water and affect local agriculture—a fact being kept hidden from the local citizenry. In an interview with him, he told me:

> I moved to Mendoza in 2015. My intention was to build a self-sustaining ecological community there. The project was going very well, but then, in 2017, fracking companies arrived in the province. A person I knew in the local government talked to me about an environmental impact study on the first four pilot fracking wells in the area. It had revealed "high levels of hydrocarbon leakage into water tables." Essentially, our water was being polluted with fossil fuels. This was being kept under complete secrecy by the government. But we knew it would be a communicational bomb if it got leaked to the public. Mendoza is highly seismic and desertic. Fracking causes both increased seismic activity and desertifies the land. On top of that, the province had previously gone through a major series of protests against megamining, which like fracking also pollutes the water. We knew we needed to call attention to the fracking report. We made a rudimentary Facebook page and started releasing the report under the name EcoLeaks. Initially, government-friendly media was deny-

ing everything, but then, once the testing labs confirmed the results, the communicational bomb truly exploded.

The leaked report mobilized tens of thousands of people, with small cities like General Alvear seeing 15,000 out in the streets (more than half of its population). Highway blockades were mounted across Mendoza, and central squares in many towns were occupied. Consequent police repression led to dozens of arrests, and the protests precipitated the drafting of seven anti-fracking law projects at the Mendoza province legislature.

Despite all the efforts Esteban made to expose the ties between the government and fossil capital, he was eventually forced to leave the land he had fought for. The Mendoza governor launched a police investigation against him and the EcoLeaks whistle-blower sources, involving heavy criminal charges. In 2019, after persistent intimidation from government employees, including multiple fabricated criminal charges and threatening phone calls, he had to seek exile in Europe.[6]

I met Esteban in a courthouse in central Copenhagen in 2023, minutes before he was scheduled to face a Danish judge. Since his exile, Esteban has not stopped fighting against fracking. In 2021, he and 20 other activists occupied the Scandinavian headquarters of TotalEnergies, one of the fossil fuel corporations operating in Argentina. The activists bypassed security and barged into lobbies, hallways, and executive offices, demanding an end to the company's abuses in the Global South. Esteban even confronted Total's regional CEO, asking him why he still stood by the destruction his company was wreaking in Argentina. Through fracking, Total and other companies are currently running another oil extraction operation in Neuquén (neighbor province to Mendoza) in the second-largest tight shale reservoir in the world—a place known as *Vaca Muerta* ("Dead Cow" in Spanish). The shale has been called a "carbon bomb": its exploitation has the potential to release up to 50 billion tons of CO_2 during its operation[7] (in comparison, the entire world emits roughly 37 billion tons per year from fossil fuels and industry). Total pressed criminal charges on Esteban for his actions. When I talked to him, he was having his day in court. Even after he met the Danish judge, he remained defiant: "I come from Argentina, where you don't have a choice to get depressed or get burnt out, because you have to fight for your survival."[8]

The fracking operations in Vaca Muerta demand huge amounts of water and are devastating agriculture in the region. The sludge waste is also polluting the land. Since 2013, the native Mapuche Indigenous communities have engaged in civil resistance, blocking oil access roads and occupying fracking towers. To this day, they continue to demand an end to fracking but have only been met with intimidatory threats and mass arrests by the local government.[9]

Esteban understands that Argentina is not unique. Countries like Uganda, Tanzania, Nigeria, Panama, and Indonesia are also facing incursions by powerful multinational corporations. As in Argentina, this is facilitated by complicity with local governments, dependent on foreign capital through the global debt system. The debt is an instrument of financial domination, which ensures Global South countries remain subservient to Global North interests: governments in the South are pressured into taking loans from North-controlled institutions like the International Monetary Fund (IMF) and the World Bank. The loans impose strict conditions on those countries that take them, ensuring that Global North companies, like Total, receive special privileges inside their territories. This debt is rarely paid by the same government that originally takes the loans, and it often requires the accrual of further debt by subsequent governments. As Esteban puts it, by keeping states under the leash of fossil corporations, the IMF and the World Bank are the "real climate criminals."

In 2020, Esteban co-founded the Debt for Climate grassroots movement, which calls for an end to this vicious cycle of domination. The movement demands the complete abolition of the debt system so that Global South countries can have the freedom to transition away from fossil fuels:

> Debt for Climate is a tool for amplifying power, for amplifying the struggles we are facing in the Global South. It's about directing the energies of

activist groups in the North in the service of struggles in the South, against extractivism and colonialism. And the unpayable debt is the common denominator of these struggles.

Thus, the strength of the movement's campaigns has come from the clever articulation between protests in the North—where the headquarters of financial institutions are located—and the South—where the consequences of their economic policies are felt. For example, in June 2022, Debt for Climate organized simultaneous protests in Europe, Latin America, and Africa. Occupations held in fossil-funding banks and other financial bodies in Germany, Spain, and the United Kingdom happened in tandem with mass street protests in Tanzania, Argentina, Mexico, and Paraguay. Led by worker and Indigenous groups, the protests highlighted the environmental abuse caused by the imposition of policies from the North, which enable multinationals to operate freely in the South.

ALL FOR ONE AND ONE FOR ALL

In the global fight against fossil multinationals, arguably the hardest battles today are being fought in Sub-Saharan Africa. As we saw in chapter 6, Total is building one of its largest projects there—the East Africa Crude Oil Pipeline, or EACOP for short. Running from Lake Albert in Uganda to the Port of Tanga in Tanzania, the pipeline is expected to be 1,443 km in length and carry 200,000 barrels of oil per day (approximately the amount of oil that the country of Finland consumes daily). Its construction is already displacing thousands of people from their homes.[10] Bultfran is an environmental scientist and SR member from Uganda, where the pipeline is meant to connect with oil refineries for transportation out of the country. He stressed to me how EACOP has caused strife in local communities:

> We are losing a lot of biodiversity in areas where they're constructing EACOP and the oil refineries. I've been to those rigs. Forests have been cleared, animals have died, people have been displaced from their heritage. In Africa, people have a strong attachment to their ancestral lands, and all this has been destroyed. There's a lot of contamination around the

lake already. In the future, we expect there to be oil spills and corrosion from the pipeline. Dangerous gases are likely to leak from it, like hydrogen sulfide.

The oil extracted from these megaprojects is expected to be shipped away, feeding the Global North's thirst for petrol at the expense of those who will suffer the environmental consequences:

> We don't know if anyone locally will actually benefit from EACOP. Oil will be pumped away into Asia or Europe. This is already a major injustice to the locals living there, even as we also recognize we need to move away from fossil fuels and think about projects that are green, like solar and wind.

But resistance is mounting. Bultfran makes it clear that many are daring to speak up against the pipeline. Scientist Rebellion Uganda has done a number of civil disobedience actions in the country to alert people to the dangers of these projects:

> We block roads and we try to stop cars. We even blocked the main Kampala-Jinja highway. Whenever we do this, we have a dialogue with police and generally try to avoid arrests, as prisons are terrible here. We have also done protests in universities, like Kyambogo and Makerere University, posting papers on walls. Once, we organized a protest in Hoima, the "oil city" of Uganda, which is close to the refineries and the start of the EACOP pipeline.

Shami is another SR activist, now living in Rwanda, but he was originally raised in Uganda. He explained that many young people back home, especially students, have faced the consequences of standing up against EACOP:

> Resistance against EACOP came through because people have been learning. They realized that what they had been doing for a long time against fossil projects had not worked. Now resistance is on. And it's coming mostly from civil society. Yesterday I was talking to somebody who is in a group called Justice Movement Uganda and is just 22 years old. He was saying that police picked him up while he was in class at Kyambogo University and threw him into jail.

The student protests against EACOP have heightened international awareness about the pipeline, evincing the state repression that Total can harness in their efforts to complete the project.[11]

As these protests were ramping up, an even bigger catastrophe was in the making just next to Uganda, in the Democratic Republic of Congo (DRC). In 2021, the DRC government announced it would auction off vast swathes of the country to Big Oil: 30 blocks of public land—covering over 11 million hectares (an area roughly the size of England)—would be sold off.[12] Much of this land overlaps with rainforests that are enormously rich in biodiversity and are home to elephants and bonobos, as well as countless other species at risk of extinction. Gérardine Deade Tanakula, an SR DRC activist, stresses that there are many communities that depend on these forests and rivers: "If the government turns them into oil blocks, that will destroy the lives of the local population. They have no one that represents them now." What's more, many of the land blocks are in tropical peatlands. These are carbon reservoirs that lock in decaying plant matter. If disturbed, they could release huge quantities of naturally stored CO_2 into the atmosphere, on top of the CO_2 released from burning oil and gas from those reserves.[13]

As in Argentina, grassroots resistance in Africa is emerging from below. Coalitions of workers, academics, students, and Indigenous communities have been making a stand against both EACOP and the DRC auctions. This struggle transcends borders: mobilizations occur across several countries, and often in coordination. Shami describes the collaborative networks that have emerged in response to these projects:

> We have managed to get at least seven Scientist Rebellion groups in Africa to do actions with a message saying "STOP EACOP AND DRC AUCTIONS." Because we are all facing the same consequences, we all recognize that fossil fuels are a threat to all of us, and that's a uniting factor to do these actions in coordination. We did a protest in Rwanda wearing lab coats right in front of the Congolese embassy, displaying a banner demanding the president of the DRC to stop the auctions. In Congo, they did a protest at the embassy of Tanzania, where EACOP ends. In Malawi, they did an action at Lilongwe University of Agriculture with the same

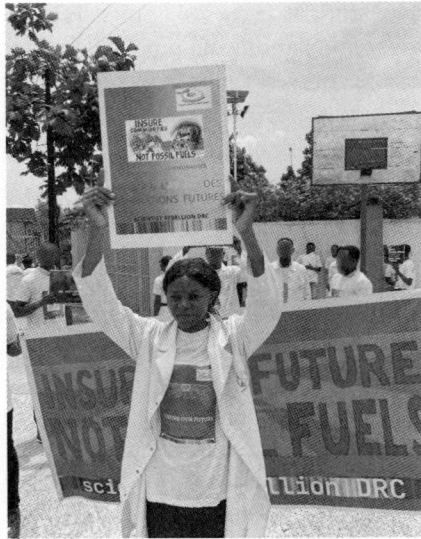

FIGURE 18. Scientist Rebellion DRC members at a protest in Kinshasa, demanding an end to investments and subsidies for fossil fuels. Credit: Jean Bondeko.

message. When we do these actions internationally, and the narrative and demands are the same, local politicians are surprised at the amount of attention we get. It helps to spread the message farther and much faster.

While the first auctions in the DRC were scheduled for January 2023, the massive pressure from activists across Africa pushed Minister of Hydrocarbons Didier Budimbu to postpone the auctions until later that year.[14] The COP28 meeting in November 2023 led to another delay, as the DRC government wanted to divert attention away from fossil extraction within its borders. Nevertheless, the auctions were expected to resume in 2024. Yet, that same year, the constant campaigning bore fruit: the DRC government finally canceled all of the oil auctions in October 2024.[15] This has alerted many in the African climate movement to the need to keep up sustained pressure through protests. As Shami puts it, "these actions have stood out as something we have to do continuously." Sustained activism in the continent has been facilitated by cross-border solidarity and rousing narratives. In order to export oil and gas into the Global North, fossil infrastructure extends across multiple countries, and so must resistance against it. Connections are made between local demands, highlighting that this is a battle everyone must fight together.

A core motivation driving the struggle against these projects is energy independence: international campaigns in which SR Africa has participated, like Don't Gas Africa, are structured around the dispossession of fossil fuels away from the continent.[16] Energy capacity remains very low in Sub-Saharan Africa, with more than 600 million people lacking access to electricity and over 80 percent of the population not able to use clean cooking technologies. Even though African soils harbor large quantities of oil and gas, projects like EACOP and the DRC auctions are a way for them to be easily ransacked and placed into global markets—this has been termed an "energy apartheid." Meanwhile, investments in sustainable technologies, like solar and wind, are discouraged by Global North companies, as they would grant local communities much more control over how energy is generated and who would get to use it.

SCIENCE IN DIALOGUE WITH INDIGENOUS VOICES

One of the main roles of Scientist Rebellion in Africa has been as an articulator of demands—a way for resistance groups to find common ground, under the banner of scientific knowledge. Scientist Rebellion is also playing this role in other countries across the Atlantic. For example, ecologist Jordan Cruz described to me how Scientist Rebellion facilitates intergroup dialogue for the environmental movement in Ecuador:

> Across our campaigns, there have been many local collectives working together with us: Extinction Rebellion, Fridays for Future, Acción Ecológica. Our idea as Scientist Rebellion has been to try to collaborate and find ways to unite the movement, rather than imposing an agenda of our own. In some sense, this is how I think science should operate in general—not as the main actor, but as a roundtable where dialogue can be

fostered among different actors and political decisions can be taken, on the basis of scientific information.

Thus, activists can use the accrued prestige of science and scientists as a way to ground their common struggles: *we are all coming from different corners of society, but we all act based on what the science is telling us.* At the same time, as Jordan stresses, scientist-activists in Ecuador are careful not to "impose their own agenda" and to be mindful of those with much more experience in environmental struggles:

> In SR Ecuador, there is something that we've always said from the beginning: we didn't invent ecological resistance. There is a much longer tradition of resistance here, mainly by Indigenous communities. In Ecuador, 13 percent of the population are Indigenous, and many more have Indigenous ties. Indigenous resistance groups here are well organized and have been successfully fighting against extractivism for a long time.

This is not just true in Ecuador. Resistance by Indigenous groups—through blockades, sabotage, and other forms of direct action—is stopping or delaying projects equivalent to at least 24 percent of the annual emissions of what is currently Canada and the United States.[17] And just south of the US border, in the Mexican states of Chihuahua and Coahuila, coalitions of Indigenous people, in alliance with women's rights groups, local scientists, social movements, and NGOs, have mounted an impressive anti-fracking campaign involving highway blockades, university protests, and legal action. The campaign, which started in 2013, led to a ban on fracking in Mexico, which was in effect as of 2021.[18]

Indeed, across the world, bottom-up mobilizations by Indigenous peoples have been highly effective at halting the advance of fossil capital.[19] This is partly because of the ties that local defenders have to the lands that these projects threaten to destroy. As we saw in chapter 4, their knowledge systems are rooted in relationality—in their connections to the land and to each other. Their actions are also grounded in these connections, making it difficult for states and corporations to nullify them. And the arguments that Indigenous defenders use to justify their resistance dig deep into the roots of injustice. Unlike other climate groups, Indigenous defenders can draw direct links between the destruction of their land and the logics that define modern society, like profit extraction and domination over the Other. In Indigenous struggles, the words *capitalism* and *colonialism* are never left unmentioned.[20] At the same time, local defenders also face the greatest amounts of repression from these systems: while protecting their lands, they have been disproportionately murdered by state and corporate actors, at particularly high rates in the last 20 years.[21] Yet, they continue mounting a formidable resistance against fossil capital, even while risking their lives.

One way in which Scientist Rebellion tries to honor Indigenous resistance is by avoiding settler colonial names for lands to which Indigenous groups have long-standing claims and connections. There are no SR USA or SR Canada chapters, but there is an SR Turtle Island chapter, which includes scientist-activists from both countries. Similarly, the network of SR chapters that stretches from Mexico all the way south to Argentina is called SR Abya Yala—a name used by many Indigenous groups to refer to the (so-called) American landmass. Jordan recalls:

> The name Abya Yala originally comes from the Indigenous Guna people in Panama. They use it to refer to their land, but it was later taken on in struggles by the Mapuches in the Southern Cone. Our logo in SR Abya Yala features a jaguar, which also has lots of connections to Indigenous communities. It's a charismatic species whose niche stretches across the continent. By conserving the jaguar, we conserve many ecosystems. The jaguar in our banners is roaring, which symbolizes both anger and the love of nature we feel in SR Abya Yala. It has served both as a marketing strategy but also an emotional bond, a way to connect ourselves with the movement. And these symbols have also helped greatly in reaching out to Indigenous groups—they show we define ourselves as anti-colonialist and that we cherish life outside the narrow conceptions of Western thought.

The jaguar is also the icon of the land affected by one of the most famous cases of Indigenous resistance against fossil fuels—the Yasuní National Park in the Ecuadorian Amazon. This park extends for 1 million hectares. It is home not just to jaguars, but also more than 500 bird species and over 100 bat species. The Yasuní is actually one of the most biodiverse places on the entire planet: just its native species of frogs and toads far outnumber those found in the entirety of the United States and Canada.[22] Under the park's luscious tropical forest lies its curse: Ecuador's largest petroleum reserve, containing an estimated 1.7 billion barrels of crude oil. For this reason, the region has been highly coveted by energy companies for decades.

The fight against fossil fuels in the Yasuní dates back to the 1990s and is among the earliest instantiations of the wider Blockadia phenomenon.[23] Here, Chevron dumped billions of gallons of toxic waste from its drilling operations, contaminating rivers, farms, and villages. This has led to an extremely high incidence of cancer and other diseases over the last 30 years. Many native residents died from drinking polluted water. But the local communities fought back, through a campaign called *Chevron Tóxico* ("Toxic Chevron"), and won a large class action lawsuit against the company. Chevron then mounted a marketing countercampaign. Jordan says the company "spent more on advertising themselves as sustainable than what they were meant to legally pay back for their pollution to the Amazonian communities."

After this campaign, Chevron eventually left Ecuador and sold their oil operations to the state-owned oil company—Petroecuador. In 2005, the Ecuadorian government launched plans for expanded oil exploration in the area. Drilling started in 2016, and by 2019 it had even expanded into areas previously considered protected, as they were the traditional homes of many Indigenous peoples. This again led to national and international campaigns to defend the park from fossil interests.

More recently, scientists under the banner of SR Ecuador have joined in the defense of the Yasuní. Jordan explained to me that—through Scientist Rebellion—they helped build an anti-extractivism coalition to protect the park. By joining forces with environmental organizations like *Yasunidos* ("Yasuní United") and several Indige-

FIGURE 19. Jordan Cruz in conversation with Waorani leader Alicia Cahuiya during the delivery of a claim of unconstitutionality of oil activity at the Constitutional Court in Quito, 2023. Credit: Karol Jaramillo.

nous peoples, including the Waorani—whose homeland is inside the park—they learned they had greater strength when coordinating their efforts. The coalition has managed to unite around 60 environmental groups fighting to preserve the Yasuní. Through scientific outreach, direct action, and institutional pressure, they launched a campaign called *Sí al Yasuní, sí a la Vida* ("Yes to the Yasuní, yes to Life").[24] In 2023, the coalition managed to score one of the biggest victories for the environmental movement anywhere in the world. For the first time in history, the Ecuadorian people voted through a nationwide referendum to enact a total ban on oil drilling in the park. Even as the country's own president and energy minister had denounced the ban, almost 59 percent of the people voted in favor of it, effectively ending decades of oil exploitation in the region.[25]

SOCIAL JUSTICE AS CLIMATE JUSTICE

A common factor of struggles against extractivism in the Global South is the prevalence of poverty, as well as the limitations this places on mobilization. As Jordan Cruz explains, "you can't ask someone to care about climate change if they are not able to satisfy their very basic needs." But this poverty does not emerge out of nowhere. It is intimately tied to the small groups of people who determine how wealth is spread around. Dr. Ornela De Gasperin Quintero, a professor in ecology and active member of Scientist Rebellion Mexico, stressed to me the other side of economic injustice—the ultrarich elites, who benefit from economic inequality yet are often overlooked as targets by scientist-activists:

> Many groups in SR Abya Yala focus on extractivism and debt cancellation, which makes a lot of sense. But I've been pushing to also attack the elites! Mexicans in particular need to massively change our discourses around inequality. Here in Mexico, we have over 8 million people in extreme poverty, and 60 million people are in need of social infrastructure and hospitals. At the same time, we have this very small niche of extreme wealth. Some areas of Mexico look like Monaco! The richest Mexican man pollutes more than several countries. We need to tax the rich out of existence, and we need popular support for that.

Elites tend to be a major roadblock to climate action, not only because of the influence they exert over the government and media, but also because they are especially well protected under the law. Elites are often enriched by partnerships with international fossil fuel and metal mining corporations. As Ornela explains, these companies have worked with their home states in the Global North to secure legal tools that prevent Global South governments from banning or even regulating their activities. Through so-called investor-state dispute settlements (ISDS), companies can literally take a state to court if the government dares to crack down on them: "Even if our government wanted to push for harder measures, then it would just be sued." In the year 2023, following the progressive reform of Mexico's energy and mining laws (giving legal priority to people and to nature over corporations), the government was subject to 10 lawsuits via the World Bank's ISDS.

Using direct action, Ornela is working in SR Mexico to highlight legal structures that maintain poverty and protect elite power. Because these structures are based on international agreements, SR carries out demonstrations and paper pasting in embassies of Global North countries, like Canada, the United Kingdom, and Australia. Additionally, to connect climate struggles with the elites, common action targets include luxury stores in the "Monaco-like" areas of Mexico City. As in other parts of Abya Yala and in Africa, alliances with other sectors of society are key. Part of the work involves evincing the links between economic and climate demands:

> The current government is doing lots of social justice reforms but not narrated through a climate narrative. They just created a public banking system—that's a climate action! They are taxing corporations for the first time in Mexican history—that's a climate action!

Thus, Ornela and her fellow SR Mexico members are using their elite-oriented actions to efface the line that separates the climate movement from social movements that tackle marginalization and the rights of workers. After all, the reforms these movements are pushing into the halls of government are also climate reforms, because they limit the power of polluting corporations. As Ornela puts it, "we need social justice to get climate justice."

THE DARK SIDE OF THE GREEN TRANSITION

Another reason why social and economic injustice can't be separated from climate justice has to do with the proposed "solutions" to climate change. Consider the "green transition" that is so heavily advertised in the rich countries of the Global North. Where will resources come from in order to build all the new solar panels and wind generators? Where will those be built, and by whom? Narratives in Europe and North America about this transition assume that the environmental footprint of the Global North can be reduced without thinking too much about these questions. The reason for this? Well, their proponents say, we can unlink economic growth from its effects on the environment. We can keep on running our economies the way we always have, by pursuing "decoupled growth"—in other words, consuming less and less materials even as our GDP continues to rise.

As we saw in chapter 1, this is a myth. A recent survey of 835 peer-reviewed studies shows that there is little to no evidence that countries in the Global North are successfully decoupling their economies, as advocates of *green* growth claim they are.[26] Behind the myth, however, lies a nasty truth. Without radical economic transformation, away from growth-hungry capitalism, the rising resource inputs of already overgrown nations will need to come from somewhere else. The minerals to build new solar panels, new wind turbines, new electric cars, and other "environmentally friendly" gadgets will not gently rain down from the air. They will have to be violently unearthed from the ground, in places where it is cheap to do so.

Like the climate emergency, this is not a question of the future alone. It is already happening. Throughout my interviews with scientist-activists operating in Abya Yala, I got the impression that resistance against metal mining was sometimes perceived to be more pressing than that against fossil fuels. Ornela mentioned silver and lithium extraction projects wreaking havoc on Mexican soils, while producing severe hydric stress and pollution in local communities. And in Ecuador, Jordan pointed to megamining projects for gold and copper in the Cordillera del Cóndor, in the southeast of the country—another region rich in biodiversity and home to Indigenous peoples. There, hundreds of local members of the Shuar communities have been violently evicted to allow companies from China and Canada to enter into the region.[27]

But perhaps nowhere is the struggle against metal extractivism better illustrated than in Panama. In October and November 2023, the country was witness to mass protests against mining for copper—an important component of electric car batteries. By the country's standards, the protests were enormous: spurred by government plans to turn Panama into a "mining country," over 30,000 people (almost 1 percent of the country's population) poured into the streets. Daily demonstrations and roadblocks extended for more than a month, effectively shutting Panama down.

One of the loudest voices of scientist-activism there is Dr. Renate Sponer: an Austrian geneticist who moved into the Panamanian countryside in 2000 and has since helped establish a local chapter of Scientist Rebellion called *Ya Es Ya!* ("Now Means Now!"):

> We are in a fight against megamining. Our struggle here is not related to fossil fuels. Our true fight is about defending our nature, which in Panama is so diverse but that is being sacrificed by the greed of our governments, who have been practically bought by mining companies.

On October 20, 2023, Panama's government signed a contract with the Canadian company First Quantum Minerals, sanctioning its operations in an open-pit copper megamine extending for 130 km², surrounded by rainforest.[28] Six years before, the Panama Supreme Court declared an earlier contract to operate the copper mine unconstitutional. Yet, in defiance of the court and the country's constitution, the company had illegally continued working the mine for years, devastating the surrounding rainforest. In the process, the mining residues were contaminating pristine rivers and streams, from which the region derives its drinking water. The new government contract was meant to provide a legal veneer to First Quantum's activities so that the company could operate unchallenged.

The protesters in the anti-mining mobilizations were highly diverse: student councils, labor unions, environmental groups, and Indigenous communities were all united under the same banner: "Panama is worth more without mining." As Renate explained to me, the reason for the protests was not just about the contract between the government and First Quantum, but also about new planned mining projects across the whole country: the government had already granted 15 pit concessions, and it was evaluating 104 additional corporate requests for more mining.

After weeks of protests, on November 28, a new Supreme Court ruling ordered the complete shutdown of the mine. This time, the ruling was backed by a moratorium on new mining concessions that had just been passed by the Panama Congress. It was a death blow to First Quantum. Protesters poured into the streets once more, now in celebration. One of the protests' spokespeople, Lilian Guevara, summarized the outcome in plain terms: "The people of Panama have decided we don't want to be a mining country."[29]

Renate was actively participating in the protests, lending a scientific voice to the cause through TV interviews and social media postings. However, she noticed that not many other scientists were present on the streets. She explains that some of this has to do with conflicts of interest—many of the science faculties in Panama harbor geology and geophysics departments that are involved in consultancy with the mining companies themselves. In contrast, she highlights, the youth organizations inside the universities were very active: "The students have done a lot to make this struggle as massive as possible." Renate also points to a lack of scientific support beyond the borders of Panama:

> I am very thankful to the global Scientist Rebellion movement for everything I have learned about activism from them and the mutual support I have gotten. But I am somewhat discouraged by the lack of interest we have gotten from Global North groups in the fight against megamining. This is what I think is missing in SR—to better establish links between renewable energy and megamining projects. In some sense, I am seeing similar patterns reproducing themselves as those that separate the Global North from the Global South. These are things we criticize but that we have not yet managed to overcome.

Here, Renate highlights the problems she encountered while trying to spread the word about the protests: even though the country was at a complete standstill, it was difficult to get scientist-activists in Europe or North America to care enough and amplify what was happening. Similar observations were made by other activists I talked to, in both Africa and Abya Yala. Often, struggles in the Global South are overlooked because they are not close to perceived centers of economic power. Sometimes, they are overlooked even when protests there involve much higher risks than in Europe or North America. In the Panamanian case, for example, two activists were killed, and a policeman shot another one in the eye.[30] In the words of Shami from SR Rwanda, "what we do in the South is very risky, and people have to pay attention to this."

OVERCOMING ARROGANT IGNORANCE

Renate's and Shami's criticisms exemplify how Scientist Rebellion can be confronted with its own contradictions and internal debates, debates about power structures that exist within the movement itself. These structures sometimes favor particular types of struggles over others and amplify the voices of scientists in the Global North at the expense of those in the South. Ultimately, like the privilege of *rather*, they reproduce privileges to which white European and North American scientists are sometimes blind. They are, however, very evident to those outside the imperial core. They are tied to the history of colonial and neocolonial exploitation that the South has suffered for centuries and which has favored some narratives and ways of understanding the world, at the expense of others. These privileges also extend to the concept of "the scientist," in both the North *and* the South. Scientist voices are often seen as more worthy of respect than other voices. Indeed, this is something that Scientist Rebellion has readily exploited: it is the reason why the media has flocked toward people wearing lab coats while getting arrested. At the same time, the elevation of scientists as *the experts* has led to the silencing of other voices—voices with copious amounts of wisdom to offer about climate and ecological struggles. In Ecuador, in Uganda, in Congo, and in Mexico, these voices include those of local and Indigenous communities that have fought against extractivism for much longer than scientist-activists have, often under harsher conditions of repression.

Decolonial scholars Rosalba Icaza and Rolando Vázquez have called out the *arrogant ignorance* of many academics in the Global

North: a way of acting as knowledge producers "that is at one and the same time pretending to be wide-ranging, or even claiming universal validity, while remaining oblivious to the epistemic diversity of the world."[31] We tend to see science—particularly science produced in imperial centers of power—as not only more important than other systems of knowledge, but even above them. Scientist Rebellion ultimately emerges from scientific institutions, and so it sometimes falls prey to arrogant ignorance as well. This can happen, as Shami and Renate argued, when a stronger focus is placed on Global North than on Global South activist struggles, diverting attention away from the colonial peripheries. In doing so, the movement diminishes the chances of achieving its own goals. Thus, many within SR are becoming more conscious of colonial modes of thinking and acting: scientist-activists working in the Global North are coming to terms with criticisms from the Global South and with their responsibility to undo structural injustices within the collective itself.

At the height of the civil rights movement in the United States, Black and women's rights activist Fannie Lou Hamer famously asserted that "nobody's free until everybody's free."[32] In the global movement against extractivism, these words remain as true as ever: we will not liberate ourselves from fossil fuels and megamining, from state and corporate repression, from colonial and capitalist domination, until everyone is free from them as well. Every fight is important, and no fighter should be left behind.

CHAPTER 11

Academia in the Age of Climate Breakdown

"One of the troubles with our culture is we do not respect
and train the imagination. It needs exercise. It needs
practice."

—Ursula K. Le Guin[1]

In early October 2022, I took a train from Copenhagen to Berlin. I was traveling together with fellow scientists I knew from Sweden, Norway, and Denmark. We were all attending a training session organized by the German chapter of Scientist Rebellion. Academics from all over Europe had been invited to attend. These were the days just before the actions in Berlin, Wolfsburg, and Munich, before Scholz's speech at the World Health Summit, before the Porsche pavilion occupation, and before Lorenzo's prison stay. We had already dipped our toes into civil disobedience in our respective countries, but these actions were planned to be much bigger than any of us had experienced before. We needed training in large-scale non-violent disruption, and so our first few days would be spent preparing ourselves for what was to come.

We arrived at a reclaimed industrial warehouse in the suburbs of the city, not really knowing what to expect. It was a cloudy fall day and the brutalist architecture around us did not inspire much comfort. Once we went inside, the mood changed: we were greeted by scientist-activists from Spain, the Netherlands, Portugal, Italy, and

France who were, like us, taking their first steps into climate activism. We were all equally apprehensive, but all happy to see fellow faces of resistance from different parts of the continent.

There is a sense of comfort that comes from meeting other activists, living hundreds—even thousands—of kilometers away, people who are fighting the same fights, people with motivations that resonate with our own. In a sense, it echoes how I feel when in an academic conference: these are "my people"—not because we have the same scientific interests, but because we share the same goals. We have gone through similar experiences, defying the conventions of our places of work or facing police when participating in disruptive protest.

During the training session, we learned skills about action planning and execution. It was facilitated by a charismatic Italian theoretical physicist—Daniele Artico—who had been through several actions with SR Germany. He gave a workshop on "de-escalation": how to calm people down when they are angry or annoyed at protesters, how to talk with them, and what type of body language to use. Daniele knew many of us were nervous about the upcoming events and wanted to make sure we were not just intellectually but also emotionally prepared for what was to come. We went through action simulations, with some of us pretending to be police, and we learned what to say (and not to say) when questioned. We also learned how to be peacefully carried away by officers while imposing maximum resistance. And it was an opportunity for us to share lessons learned from direct actions in our respective countries.

The training session was inspired by the popular "climate camps," ongoing since the early 2000s in various sites of resistance in Europe and Turtle Island: self-organized gatherings run to teach and organize for direct action against major polluters. These operate on anarchist principles: there are no leaders, and everyone contributes with whatever expertise or support they can provide.[2] Like the climate camps, our Berlin training included a lot more than just action prep. There were seminars with members of other civil disobedience groups, like Debt for Climate, as well as activists from the Global

South. We also engaged in communal activities, making new friends along the way. We helped each other cooking food, making banners, and even concocting the sticky flour mix we would later use for paper pasting. In between training sessions, we received guidance from psychologist-activists, who helped us share how we felt about climate breakdown with the rest of the group. Some scientists had brought musical instruments and played beautiful ballads during the breaks in between activities. It was a mixture of activism, science, art, community, all combined into a wonderful potpourri. And as at any other scientist gathering, many of us inevitably turned to talking about our research: "What do you do back home when you're not, umm, getting arrested?"

I couldn't help but think: *This is also what we ought to be doing in our respective countries, inside our classrooms, our offices, and our labs.* Here was a group of highly motivated specialists from different fields, coming together to share knowledge and know-how that could be put into practice for a truly transformative cause. In here, no voices were being silenced, and everyone was supporting each other. Everyone was a teacher and a student. This was more "excellent," "impactful," and "interdisciplinary" than anything I had ever experienced in any faculty committee or departmental seminar. It made me come to terms with the deep contradiction at the heart of academia today: rather than working inside university silos that are completely unsuited to the world to come, why are we not creating—together—better forms of knowledge-sharing communities? Why are we not constructing more nourishing, more caring ways to think and act, that positively affect the world outside the academy?

A RECURRING THEME

Throughout this book, I have focused on scientists who are stepping "out of the lab and into the streets." Through this lens, I have depicted the streets as the true sites of struggle. In chapter 3, I showed how this "displacement" from the traditional academic milieu generates attention from the media and the public. But at the same time, in all of the past chapters, I kept coming back to those scientists' own institutions—universities, museums, research centers, scholarly societies—as sites harboring deep contradictions tied to the climate emergency. In chapter 1, the first action under the Scientist Rebellion banner was directed at the Royal Society, calling it out for its silence. In chapter 4, Dr. Abramoff dared to protest at a scientific conference in her own field, then lost her job as a result. In chapters 2, 3, 5, and 7, we witnessed examples of scientists rebelling against a work environment that promotes submission and passivism in the face of the status quo—capitalist "solutions," worker exploitation, hypermobility, and disengagement from societal struggles. Scientist-activists stand up to their own institutions in many ways: by altering how they teach or connect with those outside the university, by disobeying laws or mandates from above. In chapters 6 and 10, we saw how fossil fuel corporations use their wealth to finance academic complicity and suppress dissent, in both the Global North and the Global South.

In all my interviews with scientist-activists, my final question was always the same: "Do you think academia (in its current form) is fit for the challenge of climate and ecological breakdown?" Not one of the responses to this question was positive. Many referred to deep feelings of disappointment, sadness, or anger at their own colleagues and mentors for not supporting or participating in activism. Some of these negative feelings extended toward Scientist Rebellion as well. Indeed, for Mike Lynch-White, the biggest criticism toward the movement he cofounded is that it has not focused *enough* on academic institutions as action targets:

> Something I think has been lost in SR is this idea that it's not supposed to be scientists half talking down to the general public, telling them that things are bad. This was supposed to be about scientists disrupting the scientific community. It was supposed to be scientists acting against other scientists—lovingly but frustratingly. The reason this has not happened is kind of obvious. It's uncomfortable to call out your own community and tell them "you're not doing enough."

Like the North-South tensions from chapter 10, this is another point of contention at the heart of SR. Should actions be focused externally, using scientists' prestige to call out governments and corporations? Or should they be focused internally, toward scientists' own institutions? If these are also heavily complicit in the climate emergency, and captured by fossil capital, why are scientists not putting more resources toward disrupting them as well? As Mike suggests, internally focused actions are a bigger ask: it takes a heavy emotional toll for scientists to confront their own colleagues, instead of politicians or corporate executives. And there is also a more material cost: in a capitalist system, academia is not just an institution for producing and sharing knowledge; it is also a means of subsistence—a source of salary. Occupying an airport tarmac or a coal mine carries at most the consequence of a fine or a stay in jail. Occupying a person's own workplace could entail the loss of that person's job.

Yet, Mike's point looms large for many a scientist-activist. With all the mobilization work SR has done, and its many individual victories, the truth is that our world continues to be dominated by fossil capital, and it is still on a path toward massively destructive levels of warming. Five years after SR's first act of rebellion, the world of tomorrow remains a terrifying prospect, and with each new catastrophe, with each new step toward authoritarianism, that world becomes more and more real. On top of that, scientists' own institutions—our universities, museums, and research centers—continue (through their actions, their partnerships, and their silence) to be complicit in the creation of that terrifying world. So what is there to do?

TOWARD A DEMOCRATIC UNIVERSITY

A possible way forward is to understand what exactly could be different in our academic institutions, if they were to actually live up to the challenge of the climate emergency. In a recent article, Callum McGeown and John Barry set out to tackle this issue.[3] Both of them are activists and political scholars, so their perspective comes as much from academic thinking as from on-the-ground praxis. They argue that academic institutions today are effectively agents of "unsustainability": in the way they are structured, they act to frustrate or delay climate action. Rather than catalyzing a transformation of our societies, they instead support the hegemony of fossil capital. We saw this play out many times throughout this book. When Dr. Hakala was trying to open up a conversation between activists and academics in her own university, the dean stepped in to shut her down. When Dr. Grimalda tried to exemplify how a long-distance research project could be carried out without stepping on a plane, he was fired. And when we, in SR Denmark, confronted the DTU director during our mock pipeline project, he admitted he was not willing to end the university's partnerships with fossil capital: it was simply too good a deal for him.

McGeown and Barry argue that, despite their current roles as supporters of the status quo, academic institutions—and in particular, universities—have the potential to be transformative, to spur change. They argue that this potential can be unleashed if they themselves undergo a transformation, so as to "explicitly and systematically reorient academic practices around social and ecological protection and priorities." For this to happen, universities must be *democratic*: their operations should be based on a collaborative dialogue and decision-making process involving everyone who makes them run. McGeown and Barry argue that a process of academic *democratization* can only come about through collective action and that activists within academia have an important role to play in this process:

> [S]uch transformations will most likely only be possible if more academics and students become willing to follow the leadership shown by groups such as Scientist Rebellion, Faculty for a Future and End Fossil, to organize and engage in radical and disruptive activism.[4]

In their efforts to organize in the interstices of the academy, several movements have already developed effective mobilization tools, which could be tailored to create change *within the academy* as well.

What are the specific structures that need to be transformed? As Dr. Hakala's standoff with her dean or DTU's partnership with Total exemplify, the university today is subject to extreme power imbalances that serve capital over people. University presidents, provosts, rectors, and heads of research centers stand above everyone else. Decision-making is concentrated in a few people—professional managers with oversized salaries who often have few incentives to do much more than maintain the status quo and cater to the desires of rich donors and politicians in office. Scholars Blanca Missé and James Martel call this a "parasitical group that was developed over time to exercise control over faculty, staff and students."[5] This overclass exists to serve corporate interests, rather than the people who study, teach, and research inside the university's walls. They also proliferate middle management positions and responsibilities, effectively converting tasks that were once collectively decided by faculty and staff into tasks that are dictated and passed down: from whoever is at the top to whoever is at the bottom. Missé and Martel argue that the work of democratizing universities must involve the abolition of these managerial positions and logics.

They also make another important point: the way the managerial overclass holds onto power is by dividing the ones below them—those with tenure from those without, those in relatively "safe" positions from those constantly worried about their employment prospects—not just precarious postdocs, researchers, and lecturers, but also cleaning, cooking, and gardening staff. This, they recognize, is a tactic of "divide and conquer" common in many other sectors of society (the health sector, the food sector, the social care sector, to name a few) where a managerial class also preys on the labor of those underneath.

Just like McGeown and Barry, Missé and Martel stress that the democratization of the university and the abolition of managerial positions can only come by "militant and collective action." To make their case, they borrow on examples of uprisings that took place in Argentina, Mexico, Brazil, and Ecuador during the twentieth century, which led to more autonomous and egalitarian forms of university governance:

> Whenever faculty have allied with students and staff, they have managed
> to radically change the governance structures. . . . The corporate model of
> governance needs to be replaced everywhere by workers' democracy and
> self-management. . . . To the question "how will the university run with-
> out its administrators?" the answer is that it will run exactly the way it
> always has, by its faculty and staff, by collective work processes of demo-
> cratic deliberation and implementation.[6]

Thus, the power to change universities lies in faculty uniting with students and precarious staff in common struggle. The abolition of the
extreme hierarchies of academia would not just contribute to the well-
being of its workers. Crucially, this transformation can enable our
universities to become fit for the social and ecological challenges we
face. As described in chapter 4, many of the arguments that scientists
give for not participating in activism revolve around fear of repercussions, as well as lack of time or lack of contact with activist groups.
Indeed, the study by Dablander and colleagues on barriers to activism
found that civil disobedience would be something a large proportion
of scientists could see themselves doing if they had the time and opportunity, and if their jobs were not as threatened by those above them.[7]

This, then, is the reason why democratic universities—free from a
managerial class—would be much more engaged universities, universities serving the needs of the societies in which they are embedded.
Such universities would be places that support and encourage active
citizenship, places where people would feel safe to advocate for the
causes they believe in. There they would not be constantly afraid of
being fired for saying or doing the wrong thing, and they would not
be overworked with tasks forced on them from above, tasks they
often find meaningless. This way, the reigning fear, domination, and
exploitation of the modern academy would be replaced by care,
autonomy, and trust.[8]

LESSONS BORROWED FROM ACTIVISM

If academics move to democratize their places of work, they must do
so from a position of humility—recognizing that there are sources of

wisdom beyond the narrow confines of the academy that can help make the academy better. In this chapter, I will try to argue that academics have a lot to learn from activists in the global climate and social justice movements. They have a lot to learn about values and principles that nurture democratic communities, that make them resilient, enjoyable, and welcoming places to be. I would like to show that these values and principles are useful not just for scientist-activism (for getting scientists to organize for change), but for making academia better *in general*.

In chapter 8, I mentioned horizontalism—the mitigation against power concentration across group structures and dynamics—as a principle that is central to many activist circles. In addition to top-down managers, a central form of power concentration in academia is that which exists in the role of the professor, relative to students. The professor dictates what the student is taught and how to teach it. The student is a passive receiver of this knowledge. There is a line separating them, and crossing this line can be seen as a transgression. This extreme difference in roles serves to foster division between the faculty and student communities, facilitating the divide-and-conquer strategy that Missé and Martel argue is favorable to top-down management. In contrast, radical pedagogist Paulo Freire envisioned an academic system that would efface the lines between students and teachers:

> Through dialogue, the teacher-of-the-students and the students-of-the-teacher cease to exist and a new term emerges: teacher-student with students-teachers. The teacher is no longer merely the-one-who-teaches, but one who is himself taught in dialogue with the students, who in turn while being taught also teach.[9]

This dissolution of the boundaries separating those who learn and those who teach resonates very much with my initial experiences in activism, when I was first transitioning from being merely a "scientist" to being a "scientist-activist." At the beginning of the book, I mentioned I started dipping my toes into civil disobedience with Extinction Rebellion. This was back in 2019, 2 years before Scientist

Rebellion carried out their first action in Denmark—the teach-in at the Climate Ministry. One of the people who welcomed me into Extinction Rebellion was a Dutch person called Tiem. He was several years younger than me, but at the time, I very much saw myself as a student of his. Tiem taught me what direct action entailed and how people prepared for it. He introduced me to other people in the movement and helped me find a role within it, a place where I could feel comfortable and use the set of skills I had at the time. He also taught me new forms of communication and community building that I had never been exposed to before. These included techniques for meeting facilitation and consensus building, which ensured that loud voices did not end up dominating a discussion. Many of the meetings I had with Tiem and other Extinction Rebellion members were in their homes and were complemented with food we were cooking ourselves. We were planning actions and building community at the same time. These new modes of acting and coacting with others would later on be instrumental in the founding of Scientist Rebellion Denmark—a project that Tiem and I decided to undertake together in 2020.

A few years later, in 2023, Tiem started a PhD program at the university where I work and ended up attending a class I was teaching on statistics and data analysis. Now, the tables were turned. In this more institutionalized setting, I was the teacher and he was the student. This generated in me a certain uneasiness because the switch in our roles exposed something uncomfortable about my own academic persona: the boundary between "Fernando the teacher" and "Tiem the student" was much more defined, more exclusionary, than the boundary between "Tiem the activist-as-teacher" and "Fernando the activist-as-student" had been. I was teaching the course material from a position of authority, from a distance; reproducing hierarchies rather than putting them into question. This experience confronted me with an internal contradiction: I had been keeping the lessons I had learned from activism discreetly boxed in and separated from the academic world. "Fernando the teacher" was not living up to the principles that "Fernando the activist" had embraced so readily, the principles I had tried to disseminate as I was trying to mobilize other scientists to join the climate movement.

Since then, I have become more aware of this contradiction, and I am struggling more actively to resolve it. I have recently started coteaching a course on degrowth and socio-ecological justice. This course was cocreated together with two fellow activist scholars—Dr. Nikoline Borgermann and Dr. Rebecca Rutt—also based in the same university as me. The course is a lot more participatory than any I have taught before: it involves a large amount of input from the students themselves. We adopt active learning pedagogies, following the footsteps of other initiatives around the world, which are reshaping the classroom in response to the climate emergency.[10] In many of the lectures, the class is taught by a guest speaker, while Nikoline, Rebecca, and I sit together with the students, participate in discussions, and ask critical questions of the lecturer. This has a double effect of emboldening the other students and allowing us to experience perspectives and vulnerabilities that would be invisibilized if we were just behaving as *the teachers*. The course includes members of activist movements and civil society groups, who co-teach the class together with the academic lecturers. All the teachers try to bring a personal perspective into each class, and they also bring food (sometimes home-baked goods or garden vegetables) to share with everyone. That way, we try to build community within the classroom. Throughout the course, we are effacing not only the boundaries between students and teachers, but also the boundaries between academics and activists, and between our academic and nonacademic lives: we are all student-teacher-activists and, well, people.

But academic behaviors are not only shaped by lessons from other activists. Scientist-activists learn a lot while engaging in civil resistance, while disobeying the law. Getting arrested—and seeing close friends and colleagues getting arrested as well—makes us better attuned to our emotions and more equipped to confront stressful situations inside the academy. Just as Teresa Santos expresses that after occupying the Volkswagen pavilion, she is now less afraid of her PhD defense, I now also feel better equipped for dealing with criticism about my actions or opinions, from my own colleagues within the academy.

And beyond changes at the individual level, social movements can also transform the way scientists interact with each other. The diversity of backgrounds in Scientist Rebellion and other emerging academic collectives means that natural scientists, social scientists, and humanists are constantly talking to each other and sharing valuable perspectives from their different angles of experience and analysis. SR member Dr. Julia Halder put it to me this way:

> In activism, we are constantly trying to understand the bigger picture. This is something that could be transferable to academia, where we're not used to "big picture" thinking. We are not used to going to the root causes of a problem, rather than tinkering away on the surface.

Activist groups not only try to get to the root of an issue, but try to create safe spaces for sharing the emotions generated by that issue: the doom and gloom but also the hopes and dreams. In activist communities, openness, support, and care take center stage. Dr. Halder continued:

> In academia, even finding one or two other people with whom to talk about the climate crisis can be a struggle. But finding someone else can give you something, and I've gained so much just from having conversations with other scientists involved in activism. Other people have also told me they've gotten a lot from a simple conversation with activists—a feeling that we are not alone.

This openness does not only extend to fellow scientist-activists. In groups like Scientist Rebellion, there is a strong effort to welcome insights and wisdom coming from outside the academy too, from environmental defenders who can provide invaluable knowledge of a kind that is rare inside university halls. As sociologist Max Koch explains, in the struggle for social and environmental justice, "the more codified knowledge of 'intellectuals' benefits more from the practical knowledge of the citizenry, rather than the other way around. The former are educated just as often as they serve as educators."[11]

THE BACKBONE OF A MOVEMENT

Scientist Rebellion has emulated many of Extinction Rebellion's communal practices of movement building. But mobilizing efforts for scientist-activists have their own specific set of challenges. Academic networks are highly international (in large part due to the widespread hypermobility practices we explored in chapter 7), and so mobilizing scientists into action involves online coordination at a greater level than in other, more place-based, movements. To achieve this, SR owes a great deal of its success to a large and dedicated web of international volunteers working on back-office tasks, in their own homes or places of work, which can be anywhere around the world. These activists take on chores that may not be as visible as blocking a road or occupying a building. They include things like organizing online talks and regeneration sessions, writing press releases for a campaign, creating engaging videos, or collecting legal fees for trials. It also involves connecting activists living in different corners of the world, who are working on common struggles. Julia Halder explains:

> The fact that we can connect people far away from each other gives us this immediate perspective on both the fault lines of climate disaster and of the headquarters of various fossil fuel companies or fossil-funding institutions. This is a huge advantage: it's a global movement, and thanks to online connections, the people affected most by climate disasters are now directly talking to people living next to those who are most responsible for these disasters, like bankers, politicians, or fossil CEOs.

Back-office work thus forms the backbone of the SR movement. It is the glue that holds activists connected with each other. Without them, SR would not be able to function. Yet, the faces and voices of back-office scientist-activists are often less visible than those of the people risking arrest. Charlie Gardner explains:

> That is a big problem for SR and the climate movement more broadly. People think the only way they can contribute is by gluing themselves on the street, and they might not want to do that. We need to make it more broadly known that there is more than getting arrested. That's hard because our exposure is tied to arrest.

In addition to making back-office work more visible, mitigating against power concentration can involve making sure that all activists—both those who are visible and those working in the background—participate on equal grounds in decision-making. Many SR chapters are now putting effort into ensuring that those taking on back-office tasks are not just delegated to a secondary role, but that they shape how the movement evolves just as much as front-line activists.

Academia also has its own backbone—all the people who make it run every day. And just as in climate activism, making this backbone more visible and more involved in decision-making could also help democratize our universities. If, as Missé and Martel propose, academics seek to abolish managerial structures and put a stop to divide-and-conquer tactics, then they must stop seeing professors as more valuable than other workers, as more worthy, relative to others. This is especially true for "elite professors" with high amounts of external funding that, under the theory of academic capitalism, play into the notion that academics are "entrepreneurs."[12] Relegated academics with little to no decision-making power include precarious researchers—postdocs and research assistants on temporary contracts—but also the janitors, the secretaries, the cafeteria workers—the ones without whom our universities would not be able to function. Organizing for a democratic academia will mean engaging in true dialogue with all these people and presenting a united front against the managerial class. This might involve facing difficult questions, particularly for those with high concentrations of power today: Are we willing to accept fair living conditions, secure jobs, and equal decision-making power for all? If so, would some roles have to disappear, or perhaps become rotational: responsibilities for all rather than privileged positions for the few? Should coordination tasks be assigned with equal frequency to those who today teach in a department, those who do research, and those who make food or clean toilets? Could we share these tasks more equally?

CONFRONTING ACADEMIC INJUSTICES

Beyond challenging hierarchies, scientist-activism can introduce broader notions of justice into the academy. Scientist-activists working together with oppressed groups realize they have enjoyed structural privileges linked to oppression. Even before the neoliberal turn, the Western model of the university was for centuries an instrument of imperialism. In North America, the original colonial colleges founded in the seventeenth and eighteenth centuries were rooted in expropriation of land and racialized exploitation of Black and Indigenous people. So were the land grants and employment practices through which the archetypal American campus was built in the nineteenth and twentieth centuries.[13] At a global scale, the Eurocentric academic system—which expanded into Asia, Oceania, Africa, and Abya Yala with the expansion of the Spanish, Portuguese, Dutch, French, and British empires—imposed a system of knowledge on Indigenous communities through violent reeducation techniques, as well as genocide.[14] For example, the prestigious Royal Society—the preeminent scientific institution that played host to Scientist Rebellion's inaugural action—was an active accomplice in the formation of the British Empire.[15] As explained in chapter 4, Indigenous modes of organizing, thinking, and relating to one another differ radically from the Western notions of detached science forced on them. Yet, as writer Peter Gelderloos shows, it is Western—not Indigenous—knowledge that has historically enabled the devastation wrought on land and peoples in the Global South:

> Ecocide is a highly technical process. Not a single one of the industries responsible for destroying our home has developed without the integral participation of academically trained experts. Not a single oil well, not a single gold mine, not a single fracking site. Scientists are shaped by the same mercenary pressures as all higher value laborers in a capitalist economy.[16]

In light of this, Gelderloos quotes Venezuelan filmmaker Adrianna Quena, who critically questions the overvaluation of academic knowledge practices, relative to those of Indigenous communities:

I think it is worth stopping to think: what kinds of questions can be posed from within that academic, scientific institution? And we should also ask, or at least doubt: how is it that the knowledge that arises exclusively from these institutions is more useful or more legitimate than the knowledge produced by people in their own territories, that have been wrapped up in absolute devastation but also intense climate resistance for years now?[17]

The places of work of many scientist-activists continue to perpetuate imperial hierarchies of domination, even to this day. In chapter 6, we saw how these hierarchies are sometimes manifested through economic domination, as when Global South universities receive money from oil companies to influence campus activities. Other times, they are manifested through active epistemic erasure: for example, when American missiles are used by Israeli troops to bomb Palestinian universities and libraries.

But there are less visible forms of coloniality underway as well. Decolonial scholar Amitav Ghosh argues that "Western intellectual and academic discourse is so configured that it is easier to talk about abstract economic systems than it is to address racism, imperialism, and the structures of organized violence that sustain global hierarchies of power."[18] Referring specifically to academics in Europe and North America, Ghosh grants that they might often be willing to reduce their carbon "footprint" but may not be so willing to reduce the power that comes with being members of the Western university. Through the prestige that the university holds, academics are able to set the agenda about what research gets done everywhere else, what types of scholarship are "valid" or "invalid," and what forms of engagement are considered "impactful." This, in turn, means that the academic system serves to silence voices that are outside of higher education institutions in the Global North—not just Global South scientists, but also working-class groups as well as Indigenous and Black communities in both the South and the North. Yet, as we saw in chapter 10, these groups and movements have invaluable wisdom to provide in the struggle for climate, ecological, and social justice: important stories to tell about how we got here and how we can move forward.

And so, democratizing academia also entails decolonizing it: deconstructing the unjust structures that enable universities to exert power over historically marginalized communities around the world.[19] This is a difficult truth to face for academics who benefit from these injustices. Yet here too, social movements can light the path and spur people to face these injustices head-on. One value that activists exercise almost daily is that of radical honesty, of speaking truth to power—even when it is uncomfortable or risky to do so— and of acting on that truth. To behave, like the scientist-activists in this book, "as if the house is on fire, because it is."[20] As Mike Lynch-White stresses, radical honesty should be especially directed toward our own communities. It can entail radical acts of resistance, like the university occupations by End Fossil from chapter 6, but it can also involve radical breaches of decorum in our everyday lives. Aaron Thierry and colleagues, for example, propose that academics center climate justice as a conversation topic in the coffee room or the classroom, even if it risks making people uncomfortable:

> In our view, a key initial step in facilitating such a transformation is to break the climate of silence on campus by making an active effort to push through the taboo and hold conversations with our colleagues about our concerns. Thinking with [feminist scholar Sara] Ahmed (2023), we have to take on a "climate killjoy" commitment: "if questioning an existing arrangement makes people unhappy, we are willing to make people unhappy."[21]

The "climate killjoy commitment" can also extend to decolonial struggles within the academy, often intimately related to climate justice struggles. Taking on this commitment may involve encouraging a fellow scientist in the Global North to reflect on the privileges that allow them to fly across the world for a conference, burning tons of CO_2 in the process, even while millions of people remain unable to even afford a plane ticket. It might also involve urging that same scientist to think about the history of imperial domination that allows their voice to be perceived as more valuable or worthy of being heard than the voices of those whose lives are being uprooted by climate change, or those who might be risking their lives to stop corporations from encroaching on their land. A democratization of academia not only would encourage academics to join the fight for climate justice, but would help them see themselves as enmeshed in the structures of power that gave rise to climate breakdown. Again quoting Paulo Freire, "a deepened consciousness of their situation leads people to apprehend that situation as an historical reality susceptible of transformation."[22]

But I need not speak in hypotheticals. Through newly built bonds of solidarity, academic activism is already bringing about a democratic and decolonial transformation within the academy. I will use a case from personal experience as illustration. A group of students at my university are seeking to get the university to divest from corporations contributing to the occupation and genocide in Palestine. Together with several academics employed at the university, I have stood in support of the students.[23] This puts us in a position of vulnerability, a position shared by many other academics around the world, who have also expressed themselves publicly in support of campus protests and encampments. But at the same time, I know that—in taking this position—there are also many scientist-activists who have my back, just as Dr. Grimalda and Dr. Abramoff knew they could count on their extrainstitutional communities to help them in times of need. This empowers me and many other scientists to stand our ground, to push the political frontier much further than if we were doing it alone.

ENVISIONING ACADEMIC FUTURES

In one of the exercises of our degrowth course, I ask the students to imagine what academia could look like in a postgrowth utopia—a world in which our societies are no longer ruled by capital accumulation and instead work toward the well-being of all. A postgrowth university, for example, could provide degrees in "exnovation"[24]— finding ways to quickly and safely get rid of fossil infrastructure. How do we dismantle an oil rig, and what could we do with its original components instead? How do we best disassemble a coal mine and turn it into a flourishing space where nature can regrow? Post-growth universities might also serve to unlearn the ideologies that support the fossil economy: Eurocentric history, anthropocentric philosophy, and neoclassical economics. These disciplines would likely not disappear as objects of inquiry, but the form of their study would change. Just as historians today seek to find the causes behind past episodes of genocide, so might future historians study these ideologies, perhaps in an effort to understand how they set the foundation for both genocide and ecocide on a planetary scale. In the future, economics departments might perhaps no longer be centered around capitalism, or even exist separate from other disciplines. New academic hubs might instead accommodate a plurality of theories and practices for organizing societies, so as to satisfy people's needs within their local social and ecological contexts.[25]

But curricular content need not be the only thing that could change in years to come. Once, when I posited the question about a post-growth university, Palesa—a student in our course—gave me an answer I did not expect: "A postgrowth future for me is a future where the university does not exist, where knowledge is not hoarded and made scarce by the enclosure of the university." At the time, her response threw me off track by its bluntness. But it is true. In many ways, the university today and throughout its history has been yet another way in which a good that should be common and plentiful is enclosed and made scarce. In this case, the good is knowledge, which should be available to all who seek it—not just those who are learned enough or wealthy enough to have access to a university education. Palesa's perspective is not uncommon. For example, Māori scholar Linda Tuhiwai Smith stresses that "for Indigenous peoples universities are regarded as rather elite institutions which reproduce themselves through various systems of privilege."[26] As an alternative, gender and women's studies scholar Paola Bacchetta and her colleagues envision a space for education that is no longer organized around hierarchies and enclosures:[27]

> The university space we want to build with you is a decolonial and abolitionist space. It works toward a world without the university. Because in

that world, there will be no capitalist mode of production—no classes, no separate education, no avowed or unavowed colonial, racist epistemologies, no meritocracy, no need to select the more "talented," no need to assign human beings to a ruling, working, or middle class. This is our horizon.[28]

The end of these barriers to knowledge might not be tragic. It might mean that knowledge becomes more accessible to those without the "right" credentials or means of access. The dissolution of Western universities might lead to new environments in which research and learning become better enmeshed with living and more conducive to joy and well-being. This is not wishful thinking. Alternatives to the Western model of the university already exist, or they are in the process of being born.[29] For example, the Universidad de la Tierra, or Unitierra, in Oaxaca, Mexico, is an institution sheltering different ways of knowledge sharing, with "no classrooms, no teachers and no formal curriculum."[30] In Unitierra, education and research are centered on dialogue and trust with local communities: its members have created a "learning commons," where people can satisfy their curiosity, learn new skills from others, and develop new ways to run their lives—all without formal student and teacher roles. In this truly Freirean institution, student-teachers decide together on how and what they want to learn.

Unitierra and similar initiatives have spurred many communities to reclaim learning, teaching, and research as common goods—instead of professionalized crafts of the elites. This drive to academic autonomy is made manifest in the Ecoversities Alliance: a network of over 200 organizations connected by the common goal of organizing knowledge in the service of local cultures and ecologies.[31] The alliance includes places advocating for radical pedagogy: they are anticolonial and follow principles of horizontalism and community building. In these environments, power is not organized through hierarchies, and so labels like "professor," "postdoc," or "provost" cease to have much meaning within them, and they might start sounding somewhat outrageous.

Now I don't fail to see the irony of advocating for the dissolution of my own profession. Most of my credentials—perhaps the reasons why you might be reading this book—are built on the Western academic tradition: I went to college and grad school, obtained a PhD, carried out research as a postdoc, and eventually got a job as tenured faculty in a Western university of the Global North. All these credentials relied on privileges that few in the world enjoy today, and they are themselves a form of privilege: they make my voice louder than that of others. They are also a way in which academia today is undemocratic: while I have climbed the academic ladder, I would much rather there be no ladder in the first place. And so do many scientist-activists I have met outside the university's walls while campaigning for change. Already back in 1980, civil rights activist and evolutionary biologist Stephen Jay Gould lamented academia's obsession with distinction and superiority, a natural consequence of institutionalized ladder thinking, of the separation between those "up high" and those "down below." Gould said, "I am, somehow, less interested in the weight and convolutions of Einstein's brain than in the near certainty that people of equal talent have lived and died in cotton fields and sweatshops."[32]

Of course, there is the very real possibility that the Western university system might simply collapse, under the weight of its own hubris and of the escalating crises it faces. It is difficult to imagine a cap-and-gown PhD-granting ceremony in a world where we might be struggling to simply grow enough food. Whether the collapse of universities as we know them means the collapse of our knowledge systems is an open question; it will depend on the resiliency of the communities that support those systems today, the people who make the academy run. And as Indigenous peoples have demonstrated time and time again, knowledge systems can persist for generations in the face of collapse. Even as I write this, after all the physical spaces of universities in Gaza have been razed, those universities still continue to operate; their lecturers continue to teach inside refugee tents; their scholars continue to deliver video seminars to international audiences; even without chalkboards, they continue to harness knowledge as a form of resistance, against all odds. This, again, points to collective organizing as key to the future. Universities as we know them today might not survive through societal breakdown, but perhaps groups of people who are passionate about research, teaching, and learning might still continue to teach, learn, and investigate the world together, in communities based on care, trust, and self-sufficiency, actively mobilizing toward a better world.

Regardless of how academia ends up changing, it is certain that it will change, as it has many times in its history. How this will occur is still a matter of debate, as is the question of whether change will come from academics themselves, or from external circumstances brought upon them. But I hope this book has made one thing clear: change is already underway. Scientist-activists are already forging ties that will prove invaluable in the years to come. There are now entire underground networks of academics connected not by common fields of study or department affiliations, but by a shared motivation to positively contribute to societal transformation. Using encrypted messaging tools, borrowing know-how from other activists, we are constantly talking and plotting: figuring out new ways to subvert academic conventions and redefining what we are supposed to be doing, and what we are not. Increasingly, we are also connecting with those outside the ivory tower, organizing more autonomously, more democratically, and cocreating new structures to help our local communities weather the present and incoming storms.

As the climate warms, we are witnessing what education scholar Peter Sutoris calls an expansion of the "horizons of the possible": the limits of what people conceive they can do together.[33] This expansion—a radical and collective empowerment—is occurring all across the world. Climate movements are joining efforts with social justice, gender rights, worker rights, and decolonial movements in the global fight against systems of oppression. Universities are no exception to this wave: scientists and scholars are recognizing they can be powerful agents of change and are embracing collective action together with these movements. How far the horizon of the possible extends depends entirely on us. We can watch the world burn, or we can help put out the flames.

A Sunny Day

I sit down on a folding chair, one of many in the camp. There's a clear blue sky above me. It is early in the morning: some students have been eating breakfast, some are basking in the sun, and some are wrestling against the urge to keep sleeping just a little bit longer. One of the students grabs a megaphone and announces the teach-in session is about to being. Slowly but surely, about 40 people start coming out of their tents and then sit in a circle around me. Over 60 of these tents now stand on the campus grounds, right next to the building where I work. I can look up and see my office floor, familiar yet distant. The ivory tower is up there, still loudly dictating what a university is supposed to be, how students and academics are supposed to behave, but now in a weaker, deflated voice. Here on the ground, something else is being born. Flags and banners are hanging from all the buildings, perched atop tall trees, and planted on the soil. What had until 3 days ago been a simple grass garden—a garden I had passed and ignored on my way to somewhere else—has now become something much more alive: a site of resistance. As the birds chirp their morning songs, the teach-in begins.

I write these words on May 8, 2024, day 3 of the Palestine divestment camp at the University of Copenhagen. The tents are all tightly packed on the Faculty of Social Sciences lawn, renamed "Rafah Garden" to remind passersby of the ongoing invasion of Rafah, the southernmost tip of the Gaza Strip. This is one of over 100 camps that have been set up all across Europe, North America, and East Asia. I am in the midst of a new wave of academic protests—propelled by students and geared toward getting universities to divest from and stop collaborating with institutions that support the occupation of Palestine. The students have invited Scientist Rebellion Denmark to give a teach-in, on the links between genocide and ecocide, on the imperial structures at the heart of climate breakdown and colonialism, and on the history of civil resistance in universities. As I open the discussion with the students, I feel as if I am sharing forbidden knowledge, knowledge that was kept from me as a student, knowledge that will come in handy to them not just in this struggle, but in many that they will face in years to come.

As I finish the teach-in, I am reminded that I too was, just a few years ago, in need of this forbidden knowledge. And just like these students, I felt both excited and scared when I received it. It was the knowledge that we could dare to think a better future is possible and that we could bring it about through collective action, through organizing, through resisting. So I would like to end this book by reflecting on the emotional journey that I have experienced since I first received that knowledge: how activism has changed my personal life and the way I relate to others.

Activism can be painful. As we saw time and time again in this book, it can involve facing violent responses from other people or the retaliation of powerful actors. Other times, activism involves confronting those around you—your closest friends, family, and colleagues—with uncomfortable truths that they might not want to hear. This is not easy. It can strain our psyche, and it can sometimes completely break up relationships. As I meet and talk with the brave students now sitting in this camp, risking arrest or expulsion from their own university, I see traces of this fear in their faces. It is a fear

I have experienced many times myself. Particularly when engaging in direct action or civil disobedience, we often fear for our jobs or our lives. Or we fear that we will hurt those close to us if we end up going to jail for a long time.

All of this takes a toll and can impact our capacity to continue demanding change. Many activists end up pushing themselves beyond their psychological limits: a phenomenon known as activist burn-out. When this happens, they lose their drive to organize for action, sometimes for months or even years. This can be a major loss to movements, particularly when those activists are taking on many roles at once. Serious efforts are made to prevent this—through communal practices that help us regain our energies after each action, rotate responsibilities, and ensure we are not pushed to a point of no return.

Despite the pain and the fear, engaging with social movements can also be highly rewarding. Activist communities have taught me kindness, selflessness, and empathy. Through them, I have met countless people who are always ready to share food, lend a hand, or provide a shoulder to cry on. They ask for nothing in return, except that one does the same with others as well. I have made many friends through my activism. Indeed, it is hard not to create deep, lifelong bonds with people with whom one has shared highly stressful situations. These new friendships have helped me leave the destructive "academic ladder" psychology behind; they have made me realize the ladder does not really lead anywhere: like many other tools of capitalism, it is just there to keep us climbing.

Activism has also been a humbling experience. Meeting activists from diverse backgrounds has helped me see myself and my academic career in a new light, recognizing privileges throughout that career to which I was previously blind. When teaching, on the first day of class, I no longer introduce myself to my students merely as an "associate professor." Instead, I make the effort to center my privileged position as a white heterosexual man from a relatively high socioeconomic background and in a relatively high position of power within the place where I work. After meeting Indigenous environmental defenders in various SR campaigns, I also recognize I don't just "come from Argentina." I have been taught the importance of stressing that I was born in what is originally Querandí land, in Abya Yala, and that my ancestors went there from Europe, as part of a settler state project. This involved the decimation of many peoples that originally cared for that land and that are still facing oppression by the state today, including Mapuches, Huarpes, Tehuelches, and Wichis.[1] And so, I think—I hope—that activism has made me humbler and more open to learning from others.

I now feel that I am less concerned with my professional growth or "development." This helps me keep my ego in check. I recognize more situations where I just need to shut up and listen, and more situations where I hold power, and where I need to relinquish it. This has opened up time and attention to more relational activities in my life: care has become more important than growth. I am less willing to spend a weekend finishing a research project, and more willing to spend it playing hide-and-seek with my son. At the same time, I feel that activism has made some conversations with loved ones more difficult than they used to be. I want to connect with them, but for this, I need them to *be there, with me*, rather than far away, in the land of business as usual. I also know that, when it comes to many injustices, I am likely still *not there* myself, still not fully aware of complicities and privileges in which I may be enmeshed.

Finally, I have come to realize that activism deeply affects how one sees the future. Many speak of climate grief: the feeling that comes with realizing—*truly* realizing—that countless have died, are dying, and will die in coming years due to climate and ecological breakdown and the consequent destabilization of our societies. This grief is painful because we know things need not be as they are. In my case, grief involves knowing that my children will grow up in a mutilated world—a world of violence and hunger, a world much more dangerous and uncertain than the one I grew up in. And so, grief is intimately tied to anger. It is an anger directed at the powerful few, but also at the many logics and structures that stand in the way of positive change. It is an anger that—like the anger of the students

FIGURE 20. The teach-in by Scientist Rebellion Denmark at the occupied Rafah Garden at the University of Copenhagen. Credit: Anna Bystrup / Studerende Mod Besættelsen.

sitting in Rafah Garden today—ultimately comes from a place of love, from a desire to make the world *better* for those alive now and for those who will come after us.

Activists often get asked whether we have hope for the future. I think hope is something we make every day—by engaging in action. There is hope, sure, as long as we continue to stand in the way of new coal mines, oil rigs, and airports and as long as we continue to say *no more* to oppression and injustice, *no more* to business as usual. But hope is created by people willing to sacrifice their time, energy, and often their freedom to challenge structures of power. And the more of us there are out there, causing trouble, the more hope there will be to share.

Acknowledgments

Writing this book has been challenging. I usually do research and teach students about how genes evolved over time—a process that can take millennia. Studying the evolution of a social movement over the course of a few years, I often found myself lacking the proper intellectual tools, wishing I had trained instead to be a social scientist or a journalist. Luckily, I had the continued support of the Scientist Rebellion movement, which helped to make up for my many failings. I could not have written this book without the aid of many scientist-activists who lent their expertise and shared their wisdom with me. Half of this book's author proceeds will therefore go to SR and other climate justice movements. I am especially indebted to all the people who agreed to be interviewed about their activism: Teresa Santos, Nana-Maria Grüning, Charlie Gardner, Tim Hewlett, Lorenzo Masini, Renate Sponer, Sanja Hakala, Anaïs Tilquin, Lou, Kip Lyall, Rose Abramoff, Jordan Andrés Cruz, Ornela De Gasparin Quintero, Gianluca Grimalda, Elena, Mike Lynch-White, Shami, Gérardine Deade Tanakula, Bultfran, Ayo Fola, Hannah, Thierry de Novadhau, Fabian Dablander, Esteban Servat, and Julia Basak Halder.

My experiences in scientist-activism have all been as part of a collective. I am grateful to all the members of that collective, as well as all the people who have supported it from the sidelines. I owe especial thanks to Tiem van der Deure, Aitzkoa Lopez de Lapuente Portilla, Laura Horn, Marta Matos,

Nikoline Borgermann, Anders Giertz, Mads Ejsing, Miren Iraeta Orbegozo, Rebecca Leight Rutt, Jens Friis Lund, Stefan Jacobsen, Tim Henke, Fátima Sánchez, Urs Schäfer, and Frederik Appel Olsen for helping me believe that it was possible to mobilize scientists into the streets in Denmark and for providing the groundwork that turned SR Denmark from a wild idea into a living, breathing movement. I am grateful to the many wonderful people I met as the SR "Øresund" group grew and as I got more involved in activism with XR, SR, and other social movements, in Denmark and Sweden. These include Adrien, Lukas B, Rita, Line, Myranda, Jess, Anna Lia, Laura KF, Simon, Ellen O, Tim W, Lenny, Clara, Lennart S, Agnete, Oscar, Lukas K, Birke, Pelle, Kalle, Mona, Emilia, Jessie, Henning, Ronan, Andreas, Dorte, Nynne, Laura BM, Tethys, Annika, Kim, Goryan, Signe, Michelle, Ane, Connie, Marie, Sylvester, Ulla, Jorge, Ida, René, Alexandra, Søren, Regi, Tanya, Luna, Sima, Nadia, Salvatore, Thibault, Denise, Jaime, and many others with whom I have shared both stressful and meaningful moments together. Additionally, I want to thank the many brave activists I have come to know (in person or online) as I interacted with SR groups from around the world, including Elia, Marta MM, Gastón, Anna, Mar, Nate, Kristiina, Jeanette, Peter, Valeria, Fernando, Elodie, Henrik, Sara, Alex, Pascal, Maria-Inti, Cornelia, Agisilaos, Fabrizio, Elodie, Joseba, Manua, Shamim, Susanne, Matthias, Cesar, Janine, Kyle, Julia, Lane, Kaïna, Ru, Aaron, Daniele, Suzanne, Tuulia, Víctor, and Audrey. You all rock.

I am grateful to my department colleagues, particularly Hannes Schroeder, Michael Borregaard, Luisa dos Santos Bay Nielsen, Anders Hansen, Enrico Cappellini, Shyam Gopalakrishnan, Jonas Geldmann, Naia Morueta-Holme, and Eline Lorenzen, as well as many others who stood up for me when others questioned or criticized my dual role as both scientist and activist. Thanks also to the past and present members of the research group with whom I work—Martin, Katia, Alba, Graham, Ioannis, Evan, Rasa, Jilong, Aleks, Anna, Elisabetta, Moi, Marida, Emma, Peter, and Victor—for bearing with me as I navigated the many complications of acting as research advisor while also engaging in activism.

Special thanks to Teresa Santos, Lou, Frederik Appel Olsen, Martin Petr, Nate Rugh, Hannes Schroeder, and Gregory Manni for reading various chapters or versions of the book and giving invaluable edits and comments. I am greatly thankful to Laura Horn and Charlie Gardner: both are constant inspirations as activists and scientists, and they educated me with a lot of the radical ideas that contributed to the pages of this book. Additionally, Hanne Strager provided lots of useful guidance in the initial stages of the writing process, while Matthias Schmelzer and Ekaterina Chertkovskaya gave very helpful advice at the time of signing the final book contract.

My wonderful editor, Chloe Layman, was supportive of this project from the moment I contacted her, and she helped give birth to it even as she was also giving birth to her own child. Chad Attenborough and Kim Robinson from UC Press gave me lots of assistance and support in the final editing stages, and my copyeditor, Lou Doucette, significantly improved the text's readability, ensuring it hit the right tones. Additionally, Oscar Berglund, John Barry, and an anonymous peer reviewer contributed with critical comments and suggestions that helped me make the text much richer and sharper than it originally was.

I would like to thank my parents—Fernando M. Racimo and Diana N. Molinario—and my siblings, Mary and Martín, for your continued family love, which now extends to Matias—born just as the idea for this book was being conceived—and to Lilly Sofia, whom we expect to arrive soon after the final draft is submitted to press. I would also like to thank Eva, who provided lots of family care (and construction!) work during the writing of this book.

To my parents in particular: you were the ones who first introduced me to the world of science, and even though you might not agree with the risks I take when engaging in activism, it is the sense of justice that you have instilled in me that pushes me to take those risks. So, as in so many other things, I blame you for this book as well. Gracias.

Finally, I would like to express my deepest gratitude to Paula Zarén. All throughout the journey of writing this book, from start to finish, you encouraged me to keep going, believing in this project even more than I did sometimes. I could not have completed it without your insightful comments, your kindness, and your love. I want to thank you for your trust and companionship both while writing the book and while mobilizing for change. Thank you. For everything.

Notes

EPIGRAPH

1. Gregory Manni, "Climate Activist's Call," 2023, https://gregorymanni
.medium.com/climate-activists-call-cdf9fd84c678

PROLOGUE

1. Mark McGrath, "Climate change: IPCC report is 'code red for humanity,'" *BBC News*, August 9, 2021, https://www.bbc.com/news/science
-environment-58130705.

2. Copernicus Climate Change Service, "Copernicus: 2022 Was a Year of Climate Extremes, with Record High Temperatures and Rising Concentrations of Greenhouse Gases," 2022, https://climate.copernicus.eu/copernicus
-2022-was-year-climate-extremes-record-high-temperatures-and-rising
-concentrations.

3. I began writing this book in 2022, and this number is already outdated. Just as I am submitting the book for publication, in November 2024, the World Meteorological Organization has announced this year's global average temperature is expected to exceed 1.5 degrees, relative to the preindustrial baseline, breaching the so-called safety limit of the Paris Agreement. World Meteorological Organization, "2024 Is on Track to Be Hottest Year on Record as Warming Temporarily Hits 1.5°C," World Meteorological

Organization, November 8, 2024, https://wmo.int/news/media-centre/2024
-track-be-hottest-year-record-warming-temporarily-hits-15degc.
    4. Jeff Tollefson, "IPCC Says Limiting Global Warming to 1.5°C Will
Require Drastic Action," *Nature* 562, no. 7726 (October 8, 2018): 172–73,
https://doi.org/10.1038/d41586-018-06876-2.
    5. Timothy M. Lenton et al., "Quantifying the Human Cost of Global
Warming," *Nature Sustainability* 6, no. 10 (October 2023): 1237–47, https://
doi.org/10.1038/s41893-023-01132-6.
    6. IPCC, "FAQ 3: How Will Climate Change Affect the Lives of Today's
Children Tomorrow, If No Immediate Action Is Taken?," 2023, https://www
.ipcc.ch/report/ar6/wg2/about/frequently-asked-questions/keyfaq3/.
    7. IPCC, *Climate Change 2022: Impacts, Adaptation, and Vulnerability.*
Contribution of Working Group II to the Sixth Assessment Report of the
Intergovernmental Panel on Climate Change. (Cambridge University Press,
2022), https://www.ipcc.ch/report/ar6/wg2/.
    8. Pablo Servigne and Raphaël Stevens, *How Everything Can Collapse: A
Manual for Our Times* (John Wiley & Sons, 2020); C.E. Richards, R.C.
Lupton, and J.M. Allwood, "Re-Framing the Threat of Global Warming: An
Empirical Causal Loop Diagram of Climate Change, Food Insecurity and
Societal Collapse," *Climatic Change* 164, no. 3 (February 19, 2021): 49,
https://doi.org/10.1007/s10584-021-02957-w.
    9. Farhana Sultana, "The Unbearable Heaviness of Climate Coloniality,"
Political Geography 99 (2022): 102638.

INTRODUCTION

    1. Robin Wall Kimmerer, *Braiding Sweetgrass: Indigenous Wisdom, Sci-
entific Knowledge and the Teachings of Plants* (Milkweed editions, 2013).
    2. Greta Thunberg, *No One Is Too Small to Make a Difference* (Penguin
Random House, 2019).
    3. Damian Carrington, "School Climate Strikes: 1.4 Million People Took
Part, Say Campaigners," *The Guardian,* March 19, 2019, sec. Environment,
https://www.theguardian.com/environment/2019/mar/19/school-climate-
strikes-more-than-1-million-took-part-say-campaigners-greta-thunberg.
    4. Matthew Taylor and Damien Gayle, "Dozens Arrested after Climate
Protest Blocks Five London Bridges," *The Observer,* November 17, 2018,
sec. Environment, https://www.theguardian.com/environment/2018/nov/17
/thousands-gather-to-block-london-bridges-in-climate-rebellion.
    5. Tom Embury-Dennis and Joe Sommerlad, "Climate Change Protesters
Are 'swarming' Major London Roads to Stop Rush Hour Traffic," *The Independent,* November 21, 2018, https://www.independent.co.uk/news/uk

/home-news/climate-change-protest-london-tower-bridge-lambeth-global-warming-extinction-rebellion-a8644166.html.

6. BBC News, "Extinction Rebellion: Climate Protesters Block Roads," BBC News, April 15, 2019, sec. London, https://www.bbc.com/news/uk-england-london-47935416.

7. Jack Hardy, "Extinction Rebellion: Climate Protesters Dodge Arrest after Police Run out of Cells," *The Telegraph,* April 16, 2019, https://www.telegraph.co.uk/news/2019/04/16/extinction-rebellion-climate-protesters-dodge-arrest-police/.

8. BBC News, "Extinction Rebellion: Arrests at Sydney and Amsterdam Protests," BBC News, October 7, 2019, sec. World, https://www.bbc.com/news/world-49959227.

9. Extinction Rebellion, *This Is Not A Drill: An Extinction Rebellion Handbook* (Penguin UK, 2019).

10. Roger Hallam, *Common Sense for the 21st Century: Only Nonviolent Rebellion Can Now Stop Climate Breakdown and Social Collapse* (Common Sense for the 21st Century, 2019).

11. Gregor Hagedorn et al., "Concerns of Young Protesters Are Justified," *Science* 364, no. 6436 (2019): 139–40, https://doi.org/10.1126/science.aax3807.

12. 174 forskere, "Åbent brev fra 174 danske forskere: vi bakker op om Extinction Rebellion," Solidaritet, May 28, 2019, https://solidaritet.dk/174-forskere-stotter-klimastrejker/.

13. The Guardian, "'We Declare Our Support for Extinction Rebellion': An Open Letter from Australia's Academics," *The Guardian,* September 19, 2019, sec. Science, https://www.theguardian.com/science/2019/sep/20/we-declare-our-support-for-extinction-rebellion-an-open-letter-from-australias-academics.

14. Scientists for XR, "Scientists for Extinction Rebellion," Scientists for Extinction Rebellion, 2023, https://www.scientistsforxr.earth.

15. Before the global mass climate protests, one of the earliest examples of a scientist openly disobeying the law to call for climate action was Dr. James Hansen, known for his 1988 US congressional testimony on climate change and more recently for his acts of civil disobedience against mountaintop coal mining in 2009 and the Keystone XL Pipeline in 2013.

16. Scientist Rebellion, "Charlie Gardner: The University in Times of Climate and Ecological Crisis," 2022, YouTube video, https://www.youtube.com/watch?v=g4CZy7BDiWs.

17. Charlie J. Gardner and Claire F. R. Wordley, "Scientists Must Act on Our Own Warnings to Humanity," *Nature Ecology & Evolution* 3, no. 9 (2019): 1271–72, https://doi.org/10.1038/s41559-019-0979-y.

18. Jake Johnson, "Dozens Arrested as Scientists Worldwide Mobilize to Demand 'Climate Revolution,'" Common Dreams, 2022, https://www.

commondreams.org/news/2022/04/07/dozens-arrested-scientists-worldwide
-mobilize-demand-climate-revolution.

19. Karin Hendrich, "Klima-Klebe-Professor zu 3600 Euro verdonnert,"
B.Z.—Die Stimme Berlins, June 20, 2023, sec. Menschen vor Gericht,
https://www.bz-berlin.de/polizei/menschen-vor-gericht/klima-klebe-professor
-zu-3600-euro-verdonnert.

20. Andrés Actis, "Científicos se enfrentan a penas de cárcel por la lucha
climática: 'Es una imagen muy incómoda para el Gobierno,'" 2023, https://
www.lapoliticaonline.com/espana/judiciales-es/cientificos-se-enfrentan-a-
penas-de-carcel-por-su-lucha-climatica-es-una-imagen-muy-incomoda-para-
el-gobierno/; Pablo Rivas, "El Estado contra la comunidad científica: arranca el
juicio contra 15 académicos por teñir de rojo el Congreso," El Salto, www
.elsaltodiario.com, 2023, https://www.elsaltodiario.com/desobediencia/estado
-comunidad-cientifica-arranca-juicio-15-academicos-tenir-rojo-congreso.

21. Moises Exposito-Alonso et al., "Genetic Diversity Loss in the Anthropocene," *Science (New York, N.Y.)* 377, no. 6613 (September 23, 2022):
1431–35, https://doi.org/10.1126/science.abn5642.

22. Paul Feyerabend, *Against Method,* Fourth Edition (Verso, 2010).

23. Oscar Berglund, "Disruptive Protest, Civil Disobedience &
Direct Action," *Politics,* 2023, https://journals.sagepub.com/doi/full/10.1177
/02633957231176999.

24. Andreas Malm, *How to Blow Up a Pipeline* (Verso, 2021). For a critique of Malm's perspective, see also Bue Rübner Hansen, "The Kaleido-
scope of Catastrophe–On the Clarities and Blind Spots of Andreas Malm,"
Viewpoint Magazine, April 14, 2021.

25. Samuel Finnerty, Jared Piazza, and Mark Levine, "Between Two
Worlds: The Scientist's Dilemma in Climate Activism" (OSF, June 17, 2024),
https://doi.org/10.31234/osf.io/75qbd.

26. Stephen John Quaye, Mahauganee Dawn Shaw, and Dominique C.
Hill, "Blending Scholar and Activist Identities: Establishing the Need for
Scholar Activism," *Journal of Diversity in Higher Education* 10, no. 4 (2017):
381–99, https://doi.org/10.1037/dhe0000060; Laurence Cox, "Scholarship
and Activism: A Social Movements Perspective," *Studies in Social Justice* 9,
no. 1 (2015): 34–53, https://doi.org/10.26522/ssj.v9i1.1153.

27. For studies exploring the scholar-activist identity, see Farzana Bashiri,
"Conceptualizing Scholar-Activism Through Scholar-Activist Accounts," in
Making Universities Matter: Collaboration, Engagement, Impact, ed. Pauline
Mattsson, Eugenia Perez Vico, and Linus Salö (Springer Nature Switzerland,
2024), 61–97, https://doi.org/10.1007/978-3-031-48799-6_4; Joan Mar-
tinez-Alier et al., "Between Activism and Science: Grassroots Concepts for
Sustainability Coined by Environmental Justice Organizations," *Journal of*

Political Ecology 21, no. 1 (2014), https://doi.org/10.2458/v21i1.21124; Scott Frickel, "Just Science? Organizing Scientist Activism in the US Environmental Justice Movement," *Science as Culture* 13, no. 4 (2004): 449–69, https://doi.org/10.1080/0950543042000311814.

CHAPTER 1

1. Paulo Freire, *Pedagogy of the Oppressed* (Penguin Random House, 2017).

2. Real Media, "Scientist Rebellion Throw Paint at the Royal Society—Real Media—The View from Below," Real Media, 2020, https://realmedia .press/scientist-rebellion-royal-society/.

3. IPCC, *Climate Change 2021: The Physical Science Basis*. Contribution of Working Group I to the Sixth Assessment Report of the Intergovernmental Panel on Climate Change. (Cambridge University Press, 2021), https:// www.ipcc.ch/report/ar6/wg1/.

4. IPCC, *Climate Change 2022: Impacts, Adaptation, and Vulnerability*. Contribution of Working Group II to the Sixth Assessment Report of the Intergovernmental Panel on Climate Change. (Cambridge University Press, 2022), https://www.ipcc.ch/report/ar6/wg2/.

5. IPCC, *Climate Change 2021: The Physical Science Basis*.

6. Climate Action Tracker, "Despite Glasgow Climate Pact 2030 Climate Target Updates Have Stalled," June 2022, https://climateactiontracker.org /publications/despite-glasgow-climate-pact-2030-climate-target-updates-have-stalled/.

7. Günter Blöschl et al., "Current European Flood-Rich Period Exceptional Compared with Past 500 Years," *Nature* 583, no. 7817 (July 2020): 560–66, https://doi.org/10.1038/s41586-020-2478-3; E. Bevacqua et al., "Higher Probability of Compound Flooding from Precipitation and Storm Surge in Europe under Anthropogenic Climate Change," *Science Advances* 5, no. 9 (September 18, 2019): eaaw5531, https://doi.org/10.1126/sciadv.aaw5531.

8. Camilo Mora et al., "Over Half of Known Human Pathogenic Diseases Can Be Aggravated by Climate Change," *Nature Climate Change* 12 (2022): 869–75, https://doi.org/10.1038/s41558-022-01426-1; Marina Romanello et al., "The 2021 Report of the Lancet Countdown on Health and Climate Change: Code Red for a Healthy Future," *The Lancet* 398, no. 10311 (October 30, 2021): 1619–62, https://doi.org/10.1016/S0140-6736(21)01787-6.

9. François-Nicolas Robinne et al., "Scientists' Warning on Extreme Wildfire Risks to Water Supply," *Hydrological Processes* 35, no. 5 (2021): e14086, https://doi.org/10.1002/hyp.14086.

10. Yanjun Wang et al., "Tens of Thousands Additional Deaths Annually in Cities of China between 1.5°C and 2.0°C Warming," *Nature Communications*

10, no. 1 (2019): 3376, https://doi.org/10.1038/s41467-019-11283-w; Jun Wang et al., "Anthropogenic Emissions and Urbanization Increase Risk of Compound Hot Extremes in Cities," *Nature Climate Change* 11, no. 12 (December 2021): 1084–89, https://doi.org/10.1038/s41558-021-01196-2.

11. IPBES, "Global Assessment Report on Biodiversity and Ecosystem Services of the Intergovernmental Science-Policy Platform on Biodiversity and Ecosystem Services" (IPBES secretariat, 2019), https://knowledge4 policy.ec.europa.eu/publication/global-assessment-report-biodiversity -ecosystem-services-intergovernmental-science-o_en.

12. Eimear Nic Lughadha et al., "Extinction Risk and Threats to Plants and Fungi," *PLANTS, PEOPLE, PLANET* 2, no. 5 (2020): 389–408, https:// doi.org/10.1002/ppp3.10146.

13. Priyadarshi R. Shukla et al., *Climate Change and Land*. An IPCC Special Report on Climate Change, Desertification, Land Degradation, Sustainable Land Management, Food Security, and Greenhouse Gas Fluxes in Terrestrial Ecosystems. (2019). Retrieved from https://www.ipcc.ch/srccl/.

14. IPCC, *Climate Change 2022: Impacts, Adaptation, and Vulnerability*.

15. Kathleen Magramo, "A Third of Pakistan Is Underwater amid Its Worst Floods in History. Here's What You Need to Know," CNN, September 2, 2022, https://www.cnn.com/2022/09/02/asia/pakistan-floods-climate -explainer-intl-hnk/index.html.

16. Damian Carrington, "Spain and Portugal Suffering Driest Climate for 1,200 Years, Research Shows," *The Guardian,* July 4, 2022, sec. Environment, https://www.theguardian.com/environment/2022/jul/04/spain-and- portugal-suffering-driest-climate-for-1200-years-research-shows.

17. Paul Hokenos, "Are Drying Rivers a Warning of Europe's Tomorrow?," BBC Future, 2022, https://www.bbc.com/future/article/20220912 -are-drying-rivers-a-warning-of-europes-tomorrow.

18. Oxfam International, "Confronting Carbon Inequality," Oxfam International, May 25, 2022, https://www.oxfam.org/en/research/confronting- carbon-inequality.

19. C-F. Schleussner et al., "Carbon Majors' Trillion Dollar Damages," Climate Analytics, November 6, 2023, https://climateanalytics.org/publications /carbon-majors-trillion-dollar-damages.

20. Andreas Malm, *Fossil Capital: The Rise of Steam Power and the Roots of Global Warming* (Verso, 2016).

21. Malm.

22. Isak Stoddard et al., "Three Decades of Climate Mitigation: Why Haven't We Bent the Global Emissions Curve?," *Annual Review of Environment and Resources* 46, no. 1 (2021): 653–89.

23. Jason Hickel, *Less Is More: How Degrowth Will Save the World* (London: Windmill Books, 2021).

24. John Barry, "A Genealogy of Economic Growth as Ideology and Cold War Core State Imperative*," *New Political Economy* 25, no. 1 (January 2, 2020): 18–29, https://doi.org/10.1080/13563467.2018.1526268.

25. International Energy Agency, "Renewables 2023: Analysis and Forecast to 2028," 2024, https://www.iea.org/reports/renewables-2023/executive-summary.

26. Sheila Jasanoff and Sang-hyun Kim, *Dreamscapes of Modernity: Sociotechnical Imaginaries and the Fabrication of Power* (University of Chicago Press, 2015); Bradon Smith, "Imagined Energy Futures in Contemporary Speculative Fictions," *Resilience: A Journal of the Environmental Humanities* 6, no. 2–3 (2019): 136–54, https://doi.org/10.5250/resilience.6.2-3.0136; Laura Horn, Ayşem Mert, and Franziska Müller, "The Realism of Our Time? Futures, Fictions, and the Mid-Century Bang," in *The Palgrave Handbook of Global Politics in the 22nd Century,* ed. Laura Horn, Ayşem Mert, and Franziska Müller (Springer International, 2023), 407–31, https://doi.org/10.1007/978-3-031-13722-8_24.

27. Sivan Kartha et al., *The Carbon Inequality Era: An Assessment of the Global Distribution of Consumption Emissions among Individuals from 1990 to 2015 and Beyond* (Oxfam, 2020), https://doi.org/10.21201/2020.6492; Dario Kenner, *Carbon Inequality: The Role of the Richest in Climate Change* (Routledge, 2019); Ulrich Brand and Markus Wissen, *The Imperial Mode of Living* (Verso, 2021).

28. Jason Hickel et al., "National Responsibility for Ecological Breakdown: A Fair-Shares Assessment of Resource Use, 1970–2017," *The Lancet Planetary Health* 6, no. 4 (April 1, 2022): e342–49, https://doi.org/10.1016/S2542-5196(22)00044-4; Tim Gore, "Confronting Carbon Inequality: Putting Climate Justice at the Heart of the COVID-19 Recovery," Oxfam, 2020.

29. Felix Creutzig et al., "The Mutual Dependence of Negative Emission Technologies and Energy Systems," *Energy & Environmental Science* 12, no. 6 (2019): 1805–17, https://doi.org/10.1039/C8EE03682A.

30. IPCC, *Climate Change 2022: Mitigation of Climate Change.* Contribution of Working Group III to the Sixth Assessment Report of the Intergovernmental Panel on Climate Change. (Cambridge University Press, 2022), https://doi.org/10.1017/9781009157926.

31. Aditi Sen and Nafkote Dabi, *Tightening the Net: Net Zero Climate Targets—Implications for Land and Food Equity* (Oxfam, 2021), https://doi.org/10.21201/2021.7796.

32. Mihrimah Ozkan et al., "Current Status and Pillars of Direct Air Capture Technologies," *iScience* 25, no. 4 (April 15, 2022): 103990, https://doi.org/10.1016/j.isci.2022.103990.

33. Reuters, "Global CCS Capacity Grew by a Third, but Much More Needed: Report," Reuters, December 1, 2020, sec. Commodities News, https://www.reuters.com/article/us-climate-change-ccs-idUSKBN28 B3SZ.

34. Reuters, "Global Carbon Emissions Rebound to Near Pre-Pandemic Levels," 2021, https://www.reuters.com/business/cop/global-carbon-emissions-rebound-near-pre-pandemic-levels-2021-11-04/

35. Rex Weyler, "The Great Carbon Capture Scam," Greenpeace International, May 16, 2023, https://www.greenpeace.org/international/story/54079 /great-carbon-capture-scam.

36. Roger Harrabin, "John Kerry: US Climate Envoy Criticised for Optimism on Clean Tech," 2021, BBC, https://www.bbc.com/news/science -environment-57135506.

37. Wim Carton et al., "Is Carbon Removal Delaying Emission Reductions?," *WIREs Climate Change* 14, no. 4 (May 16, 2023): e826, https://doi .org/10.1002/wcc.826.

38. Michael Keary, "The New Prometheans: Technological Optimism in Climate Change Mitigation Modelling," *Environmental Values* 25, no. 1 (February 1, 2016): 7–28, https://doi.org/10.3197/096327115X14497392134801.

39. Elizabeth A. Stanton, "Negishi Welfare Weights: The Mathematics of Global Inequality," October 5, 2009, https://www.sei.org/publications /negishi-welfare-weights-mathematics-global-inequality/.

40. Jag Bhalla, "We Can't Have Climate Justice Without Ending Computational Colonialism," *Current Affairs,* February 4, 2023, https://www .currentaffairs.org/2023/02/we-cant-have-climate-justice-without-ending-computational-colonialism.

41. Friederike C. Bolam et al., "How Many Bird and Mammal Extinctions Has Recent Conservation Action Prevented?," *Conservation Letters* 14, no. 1 (2021): e12762, https://doi.org/10.1111/conl.12762.

42. Charlie J. Gardner and James M. Bullock, "In the Climate Emergency, Conservation Must Become Survival Ecology," *Frontiers in Conservation Science* 2 (2021), https://www.frontiersin.org/articles/10.3389/fcosc.2021 .659912; Anthony Waldron et al., "Protecting 30% of the Planet for Nature: Costs, Benefits and Economic Implications," Campaign for Nature, 2020, https://www.campaignfornature.org/protecting-30-of-the-planet-for-nature-economic-analysis; Abel Thomas and Roshan Deshmukh, "Beard Grooming Market Size, Share & Trends | Industry Forecast—2026," Allied Market

Research, 2019, https://www.alliedmarketresearch.com/beard-grooming-market.

43. Steve Keen, "The Appallingly Bad Neoclassical Economics of Climate Change," *Globalizations* 18, no. 7 (October 3, 2021): 1149–77, https://doi.org/10.1080/14747731.2020.1807856.

44. Steve Keen, "'4°C of Global Warming Is Optimal'—Even Nobel Prize Winners Are Getting Things Catastrophically Wrong," The Conversation, November 14, 2019, http://theconversation.com/4-c-of-global-warming-is-optimal-even-nobel-prize-winners-are-getting-things-catastrophically-wrong-125802.

45. William Nordhaus, "Climate Change: The Ultimate Challenge for Economics," *American Economic Review* 109, no. 6 (June 2019): 1991–2014, https://doi.org/10.1257/aer.109.6.1991.

46. Damian Carrington, "Scientists Warn of Dire Impacts of 4°C Global Temperature Rise," *Inside Climate News* (blog), November 29, 2010, https://insideclimatenews.org/news/29112010/scientists-warn-dire-impacts-4c-global-temperature-rise/.

47. Timothy M. Lenton et al., "Climate Tipping Points—Too Risky to Bet Against," *Nature* 575, no. 7784 (November 2019): 592–95, https://doi.org/10.1038/d41586-019-03595-0.

48. Several tipping points are actually at risk of being activated already, at much less severe levels of warming than 4°C; see Timothy M. Lenton et al., "Climate Tipping Points—Too Risky to Bet Against," *Nature* 575, no. 7784 (November 2019): 592–95, https://doi.org/10.1038/d41586-019-03595-0.). There is evidence to suggest that one of these—the collapse of the West Antarctic Ice Sheet—may now be almost unavoidable, even under the most optimistic warming scenarios at the time of writing; see Kaitlin A. Naughten, Paul R. Holland, and Jan De Rydt, "Unavoidable Future Increase in West Antarctic Ice-Shelf Melting over the Twenty-First Century," *Nature Climate Change* 13, no. 11 (November 2023): 1222–28, https://doi.org/10.1038/s41558-023-01818-x.

49. Will Steffen et al., "Trajectories of the Earth System in the Anthropocene," *Proceedings of the National Academy of Sciences* 115, no. 33 (August 14, 2018): 8252–59, https://doi.org/10.1073/pnas.1810141115.

50. William Nordhaus, "Climate Change: The Ultimate Challenge for Economics."

51. Patricia E. (Ellie) Perkins, "Climate Justice, Commons, and Degrowth," *Ecological Economics* 160 (June 1, 2019): 183–90, https://doi.org/10.1016/j.ecolecon.2019.02.005.

52. David Graeber and David Wengrow, *The Dawn of Everything: A New History of Humanity* (Penguin UK, 2021); Charles C Mann, *1491:*

New Revelations of the Americas before Columbus (Knopf, 2005); Ronald Trosper, *Resilience, Reciprocity and Ecological Economics: Northwest Coast Sustainability* (Routledge, 2009), https://doi.org/10.4324/9780203881996.

53. Jason Hickel, *Less Is More: How Degrowth Will Save the World*; Matthias Schmelzer, Andrea Vetter, and Aaron Vansintjan, *The Future Is Degrowth: A Guide to a World Beyond Capitalism* (Verso, 2022).

54. Santiago Álvarez Garcia, "Sumak Kawsay o Buen Vivir Como Alternativa al Desarrollo En Ecuador" (Universidad Complutense Madrid, 2013). PDF retrieved May 16, 2021.

55. Ashish Kothari, Federico Demaria, and Alberto Acosta, "Buen Vivir, Degrowth and Ecological Swaraj: Alternatives to Sustainable Development and the Green Economy," *Development* 57, no. 3 (December 1, 2014): 362–75, https://doi.org/10.1057/dev.2015.24.

56. Murray Bookchin, *The Ecology of Freedom: The Emergence and Dissolution of Hierarchy* (AK Press, 1982); Murray Bookchin, *Social Ecology and Communalism* (AK Press, 2007).

57. Kate Raworth, *Doughnut Economics: Seven Ways to Think Like a 21st-Century Economist* (Chelsea Green, 2017).

58. Nancy Fraser, *Cannibal Capitalism: How Our System Is Devouring Democracy, Care, and the Planet and What We Can Do About It* (Verso, 2022).

59. Amitav Ghosh, *The Nutmeg's Curse: Parables for a Planet in Crisis* (University of Chicago Press, 2021).

60. World Economic Forum, SAP, and Qualtrics, "The Climate Progress Survey: Business & Consumer Worries & Hopes. A Global Study of Public Opinion," 2021, https://www3.weforum.org/docs/SAP_WEF_Sustainability_Report.pdf.

61. Greta Thunberg, "'Our House Is on Fire': Greta Thunberg, 16, Urges Leaders to Act on Climate," January 25, 2019, The Guardian, https://www.theguardian.com/environment/2019/jan/25/our-house-is-on-fire-greta-thunberg16-urges-leaders-to-act-on-climate.

62. European Research Council, "ERC Work Programme 2023," n.d., https://ec.europa.eu/info/funding-tenders/opportunities/docs/2021-2027/horizon/wp-call/2023/wp_horizon-erc-2023_en.pdf.

63. Willem Schinkel, "In Praise of Flatness. On Campus Protest and Academic Community," Erasmus Magazine, February 22, 2023, https://www.erasmusmagazine.nl/en/2023/02/22/in-praise-of-flatness-on-campus-protest-and-academic-community/.

64. Stuart Capstick et al., "Civil Disobedience by Scientists Helps Press for Urgent Climate Action," *Nature Climate Change* 12 (2022): 773–74, https://doi.org/10.1038/s41558-022-01461-y.

CHAPTER 2

1. Henry David Thoreau, *Civil Disobedience* (Empire Books, 2011).

2. Extinction Rebellion, "Global Newsletter #50: The Scientists Have Had Enough," Extinction Rebellion, 2021, https://rebellion.global/blog/2021/04/13/global-newsletter-50/.

3. Extinction Rebellion.

4. Extinction Rebellion; Real Media, "Global Scientist Rebellion—Civil Disobedience at Downing Street," 2021, YouTube video, https://www.youtube.com/watch?v=ObN8ZHrBiYM.

5. Real Media, "Scientist Rebellion Civil Disobedience at News UK," 2021, YouTube video, https://www.youtube.com/watch?v=mcuU_mMaiFo.

6. Elfredah Kevin-Alerechi, "Scientists Block Scotland Bridge to Protest 'Failed' COP26 Conference," Earth Journalism Network, November 7, 2021, https://earthjournalism.net/stories/scientists-block-scotland-bridge-to-protest-failed-cop26-conference.

7. Kevin-Alerechi; Tosin Thompson, "Scientist Rebellion: Researchers Join Protesters at COP26," *Nature* 599, no. 7885 (November 12, 2021): 357, https://doi.org/10.1038/d41586-021-03430-5.

8. Dana H. Taplin and Heléne Clark, "Theory of Change Basics: A Primer on Theory of Change" (ActKnowledge, 2012), https://www.theoryofchange.org/wp-content/uploads/toco_library/pdf/ToCBasics.pdf; Oscar Berglund and Daniel Schmidt, "A Theory of Change: The Civil Resistance Model," in *Extinction Rebellion and Climate Change Activism: Breaking the Law to Change the World*, ed. Oscar Berglund and Daniel Schmidt (Springer International, 2020), 79–95, https://doi.org/10.1007/978-3-030-48359-3_6.

9. Jem Bendell and Rupert Read, *Deep Adaptation: Navigating the Realities of Climate Chaos* (John Wiley & Sons, 2021).

10. Scientist Rebellion, "Charlie Gardner: The University in Times of Climate and Ecological Crisis," 2022, YouTube video, https://www.youtube.com/watch?v=g4CZy7BDiWs.

11. Charlie J. Gardner et al., "From Publications to Public Actions: The Role of Universities in Facilitating Academic Advocacy and Activism in the Climate and Ecological Emergency," *Frontiers in Sustainability* 2 (2021), https://www.frontiersin.org/article/10.3389/frsus.2021.679019.

12. Bendell and Read, *Deep Adaptation*.

13. European Investment Bank, "Majority of People in the European Union and beyond Say the Current Crisis Should Accelerate the Green Transition," European Investment Bank, 2022, https://www.eib.org/en/press/all/2022-445-majority-of-people-in-the-european-union-and-beyond-say-the-current-crisis-should-accelerate-the-green-transition.

14. Monika Pronczuk and Catrin Einhorn, "After a Bitter Fight, European Lawmakers Pass a Bill to Repair Nature," *The New York Times,* July 12, 2023, sec. Climate, https://www.nytimes.com/2023/07/12/climate/europe-nature-restoration-law.html.

15. European Commission, "The EU Nature Restoration Law," 2022, https://environment.ec.europa.eu/topics/nature-and-biodiversity/nature-restoration-law_en.

16. European Parliament News, "Nature Restoration: Parliament Adopts Law to Restore 20% of EU's Land and Sea," February 27, 2024, https://www.europarl.europa.eu/news/en/press-room/20240223IPR18078/nature-restoration-parliament-adopts-law-to-restore-20-of-eu-s-land-and-sea.

17. Municipal Waste Europe, "The EU Legislative Process," 2024, https://www.municipalwasteeurope.eu/eu-legislative-process.

18. Mark D.A. Rounsevell et al., "A Biodiversity Target Based on Species Extinctions," *Science* 368, no. 6496 (June 12, 2020): 1193–95, https://doi.org/10.1126/science.aba6592.

19. Axel Hochkirch et al., "A Multi-Taxon Analysis of European Red Lists Reveals Major Threats to Biodiversity," *PLOS One* 18, no. 11 (November 8, 2023): e0293083, https://doi.org/10.1371/journal.pone.0293083.

20. Gerardo Ceballos et al., "Accelerated Modern Human-Induced Species Losses: Entering the Sixth Mass Extinction," *Science Advances* 1, no. 5 (2015): e1400253.

21. Gardner et al., "From Publications to Public Actions."

22. Gardner et al.; Open Secrets, "Oil and Gas: Summary.," Open Secrets, 2023, https://www.opensecrets.org/industries.

23. Adam Lucas, "Investigating Networks of Corporate Influence on Government Decision-Making: The Case of Australia's Climate Change and Energy Policies," *Energy Research & Social Science* 81 (November 1, 2021): 102271, https://doi.org/10.1016/j.erss.2021.102271.

24. Alix Dietzel, "COP27: How the Fossil Fuel Lobby Crowded Out Calls for Climate Justice," The Conversation, November 21, 2022, http://theconversation.com/cop27-how-the-fossil-fuel-lobby-crowded-out-calls-for-climate-justice-195041; The Economist, "What Is the Fossil-Fuel Industry Doing at COP27?," *The Economist,* 2022, https://www.economist.com/the-economist-explains/2022/11/17/what-is-the-fossil-fuel-industry-doing-at-cop27; Adam Morton, "Australia's New Approach Was a Rare Positive at Cop27—but Now the Need for Action Is All the More Acute," *The Guardian,* November 21, 2022, sec. Environment, https://www.theguardian.com/environment/2022/nov/21/australia-cop27-climate-summit.

25. Neesha Salian, "COP28 UAE: Dr Sultan Al Jaber Named President-Designate," Gulf Business, January 13, 2023, https://gulfbusiness.com/cop28-uae-dr-sultan-al-jaber-is-pres-designate/.

26. Damian Carrington and Ben Stockton, "Cop28 President Says There Is 'No Science' behind Demands for Phase-out of Fossil Fuels," The Guardian, December 3, 2023, sec. Environment, https://www.theguardian.com/environment/2023/dec/03/back-into-caves-cop28-president-dismisses-phase-out-of-fossil-fuels.

27. Al Jazeera, "COP29 Host Azerbaijan Brands Oil and Gas 'Gift from God,'" Al Jazeera, November 12, 2024, https://www.aljazeera.com/news/2024/11/12/cop29-host-azerbaijan-brands-oil-and-gas-gift-from.

28. Justin Rowlatt, "COP29 Chief Exec Filmed Promoting Fossil Fuel Deals," BBC, November 8, 2024, https://www.bbc.com/news/articles/crmzvdn9e180.

29. Fiona Harvey, "IPCC Steps up Warning on Climate Tipping Points in Leaked Draft Report," The Guardian, June 23, 2021, sec. Environment, https://www.theguardian.com/environment/2021/jun/23/climate-change-dangerous-thresholds-un-report.

30. Juan Bordera, "How the Corporate Interests and Political Elites Watered Down the World's Most Important Climate Report," April 27, 2022, https://mronline.org/2022/04/27/how-the-corporate-interests-and-political-elites-watered-down-the-worlds-most-important-climate-report/.

31. Bordera.

32. Stephan Lewandowsky, "Climate Change Disinformation and How to Combat It," Annual Review of Public Health 42, no. 1 (2021): 1–21, https://doi.org/10.1146/annurev-publhealth-090419-102409.

33. G. Supran, S. Rahmstorf, and N. Oreskes, "Assessing ExxonMobil's Global Warming Projections," Science 379, no. 6628 (January 13, 2023): eabk0063, https://doi.org/10.1126/science.abk0063.

34. Naomi Oreskes and Erik M. Conway, Merchants of Doubt: How a Handful of Scientists Obscured the Truth on Issues from Tobacco Smoke to Global Warming (Bloomsbury USA, 2011); Riley E. Dunlap and Peter J. Jacques, "Climate Change Denial Books and Conservative Think Tanks: Exploring the Connection," American Behavioral Scientist 57, no. 6 (June 1, 2013): 699–731, https://doi.org/10.1177/0002764213477096; Peter J. Jacques, Riley E. Dunlap, and Mark Freeman, "The Organisation of Denial: Conservative Think Tanks and Environmental Scepticism," Environmental Politics 17, no. 3 (June 1, 2008): 349–85, https://doi.org/10.1080/09644010802055576.

35. William F. Lamb et al., "Discourses of Climate Delay," Global Sustainability 3 (ed 2020), https://doi.org/10.1017/sus.2020.13.

36. Julia Steinberger, "A Postmortem for Survival: On Science, Failure and Action on Climate Change," *Age of Awareness* (blog), May 8, 2019, https:// medium.com/age-of-awareness/a-postmortem-for-survival-on-science-failure-and-action-on-climate-change-35636c799971e.

37. Camille Hazard, "Une scientifique du Giec arrêtée en Suisse après une action de désobéissance civile," parismatch.com, October 11, 2022, https:// www.parismatch.com/actu/environnement/une-scientifique-du-giec-arretee -en-suisse-apres-une-action-de-desobeissance-civile-217345.

38. Antonio Gramsci, *Selections from the Prison Notebooks* (International Publishers, 1971).

39. Voltaire, *Candide* (Project Gutemberg, 2006), https://www.gutenberg .org/files/19942/19942-h/19942-h.htm.

40. Edward S. Herman and Noam Chomsky, *Manufacturing Consent: The Political Economy of the Mass Media*, Reprint edition (Pantheon, 2002).

41. Daniel Nyberg, Christopher Wright, and Vanessa Bowden, *Organising Responses to Climate Change: The Politics of Mitigation, Adaptation and Suffering* (Cambridge University Press, 2022).

42. Culture Unstained, "Sponsorship: Branding Culture from the UK to Russia," *Culture Unstained* (blog), January 12, 2018, https://cultureunstained .org/crudeconnections-sponsorship/.

43. Nadia Khomami, "Climate and Heritage Experts Call on British Museum to End BP Sponsorship," *The Guardian,* April 19, 2022, sec. Culture, https://www.theguardian.com/culture/2022/apr/19/climate-and-heritage -experts-call-on-british-museum-to-end-bp-sponsorship.

44. Culture Unstained, "Oil Sponsorship of the Science Museum," *Culture Unstained* (blog), February 23, 2021, https://cultureunstained.org/oil -sponsorship-of-the-science-museum/.

45. Terry Macalister, "Shell Sought to Influence Direction of Science Museum Climate Programme," *The Guardian,* May 31, 2015, sec. Business, https://www.theguardian.com/business/2015/may/31/shell-sought-influence -direction-science-museum-climate-programme.

46. Culture Unstained, "Artists and Visitors Shun Science Museum Event over 'Indefensible' Coal Sponsor," *Culture Unstained* (blog), September 6, 2022, https://cultureunstained.org/2022/09/06/artists-and-visitors-shun-science-museum-event-over-indefensible-coal-sponsor/.

47. Nyberg, Wright, and Bowden, *Organising Responses to Climate Change;* Ernesto Laclau and Chantal Mouffe, *Hegemony and Socialist Strategy* (Verso, 1985).

48. "Suella Braverman Tells Police to Be Firmer with 'extremist' Protesters," BBC News, November 9, 2022, sec. UK Politics, https://www.bbc.com /news/uk-politics-63573956.

49. J. L. Ferrer and Efe, "La Fiscalía pasa a considerar a 'Extinction Rebellion' y Futuro Vegetal como grupos 'terroristas,'" elperiodico, September 9, 2023, https://www.elperiodico.com/es/medio-ambiente/20230909/fiscalia -pasa-considerar-extinction-rebellion-91884518.

50. Laclau and Mouffe, *Hegemony and Socialist Strategy.*

51. See also Berglund and Schmidt, "A Theory of Change."

52. Nyberg, Wright, and Bowden, *Organising Responses to Climate Change.*

53. Culture Unstained, "Culture Unstained," Culture Unstained, 2024, https://cultureunstained.org/.

54. Esther Addley, "British Museum Ends BP Sponsorship Deal after 27 Years," *The Guardian,* June 2, 2023, sec. Culture, https://www.theguardian .com/culture/2023/jun/02/british-museum-ends-bp-sponsorship-deal-after -27-years.

55. BP or not BP?, "Sneaking a Trojan Horse into the British Museum," 2020, YouTube video, https://www.youtube.com/watch?v=ZCrd71Awe-U.

56. Tommy Vickerstaff, "To Win or Not to Win: What We Can Learn from BP Or Not BP?'s Win at the British Museum," 350, June 9, 2023, https://350 .org/to-win-or-not-to-win-what-we-can-learn-from-bp-or-not-bps-win-at-the -british-museum/.

57. Scientists for XR, "News," Scientists for Extinction Rebellion, 2024, https://www.scientistsforxr.earth/news.

58. Health for Extinction Rebellion, "1st February 2024: HXR Protest with XR Scientists and Greta Thunberg at Science Museum Debate," Health for XR, 2024, https://healthforxr.com/1st-february-2024-hxr-join-xr-scientists -and-greta-thunberg-at-science-museum-debate/.

59. Extinction Rebellion, "Young People and Scientists Occupy New Coal-Sponsored Science Museum Gallery, Joined by Broadcaster and Wildlife Campaigner Chris Packham," *Extinction Rebellion UK* (blog), April 12, 2024, https://extinctionrebellion.uk/2024/04/12/young-people-and-scientists- occupy-new-coal-sponsored-science-museum-gallery-joined-by-broadcaster -and-wildlife-campaigner-chris-packham/.

60. Culture Unstained, "Ethics Complaint Made against Science Museum Director over Adani Deal," *Culture Unstained* (blog), December 5, 2023, https://cultureunstained.org/2023/12/05/ethics-complaint-made-against -science-museum-director-over-adani-sponsorship/.

61. Matthew Taylor, "Hundreds of Teachers Boycott Science Museum Show over Adani Sponsorship," *The Guardian,* July 15, 2022, sec. Culture, https://www.theguardian.com/culture/2022/jul/15/hundreds-of-teachers- boycott-science-museum-over-adani-sponsorship; Culture Unstained, "Open Letter to the Science Museum," *Culture Unstained* (blog), November 19, 2021, https://cultureunstained.org/open-letter-to-the-science-museum/.

62. Culture Unstained, "Science Museum Group's Partnerships with Adani, Shell, BP & Equinor," *Culture Unstained* (blog), June 12, 2019, https://cultureunstained.org/sciencemuseum/; Culture Unstained, "Science Museum Cancels Event as Speakers Withdraw over Fossil Fuel Sponsorship," *Culture Unstained* (blog), April 27, 2022, https://cultureunstained.org/2022/04/27/science-museum-cancels-event-as-speakers-withdraw-over-fossil-fuel-sponsorship/.

63. Mary Archer, "Statement from Dame Mary Archer," Science Museum Group Blog, October 30, 2021, https://blog.sciencemuseumgroup.org.uk/statement-from-dame-mary-archer/.

64. World Economic Forum, SAP, and Qualtrics, "The Climate Progress Survey: Business & Consumer Worries & Hopes. A Global Study of Public Opinion," 2021, https://www3.weforum.org/docs/SAP_WEF_Sustainability_Report.pdf.

65. See Stefan Collini, *What Are Universities For?* (Penguin UK, 2012), for a critique of this ideal.

66. Nyberg, Wright, and Bowden, *Organising Responses to Climate Change*.

67. Kevin-Alerechi, "Scientists Block Scotland Bridge to Protest 'Failed' COP26 Conference."

CHAPTER 3

1. Anna Trads Viemose, "Gaden var klasselokale, cykelstien var tavle—Uniavisen," *Uniavisen* (blog), October 26, 2021, https://uniavisen.dk/gaden-var-klasselokale-cykelstien-var-tavle/.

2. Jocelyn Timperley, "The Law That Could Make Climate Change Illegal," BBC Future Planet, 2020, https://www.bbc.com/future/article/20200706-the-law-that-could-make-climate-change-illegal.

3. As I describe in the Prologue, when we blocked the road by the ministry again just a few months later, the police did not hesitate to carry out arrests.

4. Tanya Maria Møller Knudsen, "Forskere bliver klimaaktivister: Hvis bare vores forskning blev taget seriøst, kunne vi nøjes med den," Information, October 25, 2021, https://www.information.dk/indland/2021/10/forskere-klimaaktivister-bare-vores-forskning-taget-serioest-noejes; Arbejderen, "Forskere i Blokade Foran Klimaministeriet," Arbejderen, 2021, https://arbejderen.dk/indland/forskere-i-blokade-foran-klimaministeriet/.

5. Jakob Slyngborg Trolle, "Forskere blokerer gaden i protest mod dansk klimapolitik," DR, October 25, 2021, https://www.dr.dk/nyheder/indland/forskere-blokerer-gaden-i-protest-mod-dansk-klimapolitik; Niklas Asp Nielsen and Frederik Thymark, "Forskere blokerer gaden foran Klimaministeriet: »Lyt

til videnskaben!«," October 25, 2021, https://videnskab.dk/kultur-samfund
/forskere-blokerer-gaden-foran-klimaministeriet-lyt-videnskaben.

6. Viemose, "Gaden var klasselokale, cykelstien var tavle—Uniavisen";
Nielsen and Thymark, "Forskere blokerer gaden foran Klimaministeriet."

7. Brian Weichardt, "Frie Grønnes leder: Følg os eller du har en planet i
flammer," Ekstra Bladet, October 25, 2021, https://ekstrabladet.dk/nyheder
/politik/danskpolitik/frie-groennes-leder-foelg-os-eller-du-har-en-planet-i-
flammer/8928019.

8. The thesis chapter is published as an article in Danish, and the quoted
English text is available from the translation available in the thesis.

9. Frederik Appel Olsen, "Forskeroprør Ved Klimaministeriet: Urolig
Videnskabelig Etos Ved Scientist Rebellions Teach-in-Demonstration," Rhe-
torica Scandinavica 27, no. 86 (2023): 160–79.

10. Nielsen and Thymark, "Forskere blokerer gaden foran Klimaministeriet."

11. Extinction Rebellion NYC, "US Scientists Engage in Civil Disobedience to Call for a Climate Revolution," Extinction Rebellion NYC, April 7,
2022, https://www.xrebellion.nyc/news/april6-scientistrebellion; Rebecca
Newlands, "COP26 Glasgow: Scientist Rebellion Stage Another Protest,"
Glasgow Times, 2021, https://www.glasgowtimes.co.uk/news/19702787
.cop26-glasgow-scientist-rebellion-stage-another-protest/.

12. Scientist Rebellion France and Extinction Rebellion France, "Des scientifiques en procès pour avoir alerté sur la crise écologique," 2023, https://
scientifiquesenrebellion.fr/textes/presse/des-scientifiques-en-proces-pour
-avoir-alerte/.

13. XR France, "Live: Nuit de La Rébellion Scientifique," Facebook,
2022, https://www.facebook.com/xrfrance/videos/513667710244737.

14. Rachel Carson, Silent Spring (Houghton Mifflin, 1962).

15. While Carson's work was pivotal in challenging the agrichemical
industry, it was not without its own problematic blind spots. Patrick Bresnihan and Naomi Millner point out how her nostalgic imagining of small-
town America excludes particular actors that are often made invisible in
white liberal portrayals of nature and in struggles against corporate abuses:
"the settled, agrarian landscapes that recur in Silent Spring . . . reflect a
white-settled vision of America that excludes the exploitation of migrant
labour, Indigenous dispossession, and the many struggles undertaken to
resist these ongoing injustices."(Patrick Bresnihan and Naomi Millner, All
We Want Is the Earth: Land, Labour and Movements Beyond Environmentalism, First Edition (Bristol University Press, 2023)).

16. See, for example, Michael E. Soulé, "What Is Conservation Biology?,"
BioScience 35, no. 11 (1985): 727–34, https://doi.org/10.2307/1310054;

Curt Meine, Michael Soulé, and Reed F. Noss, "'A Mission-Driven Discipline': The Growth of Conservation Biology," *Conservation Biology* 20, no. 3 (2006): 631–51, https://doi.org/10.1111/j.1523-1739.2006.00449.x.

17. Collectif, "L'appel de 1 000 Scientifiques : « Face à La Crise Écologique, La Rébellion Est Nécessaire »," *Le Monde*, 2020, https://www.lemonde.fr /idees/article/2020/02/20/l-appel-de-1-000-scientifiques-face-a-la-crise-ecologique-la-rebellion-est-necessaire_6030145_3232.html.

18. Scientist Rebellion France and Extinction Rebellion France, "Des scientifiques en procès pour avoir alerté sur la crise écologique," 2023, https:// scientifiquesenrebellion.fr/textes/presse/des-scientifiques-en-proces-pour -avoir-alerte/.

19. Kelvin Mason, "Academics and Social Movements: Knowing Our Place, Making Our Space," *ACME: An International Journal for Critical Geographies* 12, no. 1 (2013): 23–43, https://doi.org/10.14288/acme.v12i1 .950.

20. Tucker Spencer, *Encyclopedia of the Vietnam War: A Political, Social, and Military History. Volume Two* (ABC-CLIO, 2011).

21. Marshall Sahlins, "The Teach-Ins: Anti-War Protest in the Old Stoned Age," *Anthropology Today* 25, no. 1 (2009): 3–5.

22. Sahlins.

23. Sahlins.

24. Charles DeBenedetti and Charles Chatfield, *An American Ordeal: The Antiwar Movement of the Vietnam Era* (Syracuse University Press, 1990).

25. Jonathan Allen, *March 4: Scientists, Students, and Society* (The MIT Press, 2019), https://doi.org/10.7551/mitpress/11671.001.0001.

26. Union of Concerned Scientists, "1968 MIT Faculty Statement" Union of Concerned Scientists," 1968, https://www.ucsusa.org/about/history /founding-document-1968-mit-faculty-statement.

27. Sigrid Schmalzer, Daniel S. Chard, and Alyssa Botelho, *Science for the People: Documents from America's Movement of Radical Scientists* (University of Massachusetts Press, 2018), https://muse.jhu.edu/book/59565.

28. Steven Rose, "Stephen Jay Gould," *The Guardian*, May 22, 2002, sec. Books, https://www.theguardian.com/education/2002/may/22/medicalscience .internationaleducationnews; SftP Editorial Collective, "Richard Lewontin (1929–2021): A Scientist for the People," *Science for the People Magazine* (blog), July 21, 2021, https://magazine.scienceforthepeople.org/lewontin -special-issue/richard-lewontin-1929-2021-a-scientist-for-the-people/.

29. Jeffrey Mervis, "As Scientists Prepare to March, Science for the People Reboots," *Science*, 2017, https://www.science.org/content/article/scientists-prepare-march-science-people-reboots; Nicholas St Fleur, "Scientists, Feeling Under Siege, March Against Trump Policies," *The New York Times*, April 22,

2017, sec. Science, https://www.nytimes.com/2017/04/22/science/march-for-science.html.

30. Scientist Rebellion, "Week-Long Scientific Resistance to Government and Industry Inaction on the Climate Emergency," Scientist Rebellion, 2022, http://scientistrebellion.org/about-us/press/week-long-scientific-resistance-to-government-and-industry-inaction-on-the-climate-emergency/; Fernando Racimo, "Scientist Rebellion: A Teach-in on the Climate and Ecological Crisis," 2021, YouTube video, https://www.youtube.com/watch?v=owHfSHB-Ias.

31. Fernando Racimo et al., "The Biospheric Emergency Calls for Scientists to Change Tactics," *eLife* 11 (November 7, 2022): e83292, https://doi.org/10.7554/eLife.83292.

32. Paulo Freire, *Pedagogy of the Oppressed* (Penguin Random House, 2017).

33. Joyce E. Canaan, "Sand in the Machine: Encouraging Academic Activism with Sociology HE Students Today," in *New Social Connections*, ed. Judith Burnett, Syd Jeffers, and Graham Thomas (Palgrave Macmillan UK, 2010), 204–32, https://doi.org/10.1057/9780230274877_13.

34. bell hooks, *Teaching to Transgress: Education as the Practice of Freedom*, First Edition (Routledge, 1994).

35. Julia Steinberger, "The Kids Are Not OK, " Yale Climate Connections, May 15, 2022, http://yaleclimateconnections.org/2022/05/the-kids-are-not-ok/.

36. Aaron Thierry et al., "'No Research on a Dead Planet': Preserving the Socio-Ecological Conditions for Academia," *Frontiers in Education* 8 (2023), https://www.frontiersin.org/articles/10.3389/feduc.2023.1237076.

37. Pam Houston, "The Truest Eye," Oprah.com, 2003, https://www.oprah.com/omagazine/toni-morrison-talks-love.

38. Boaventura de Sousa Santos, *The End of the Cognitive Empire: The Coming of Age of Epistemologies of the South* (Duke University Press, 2018).

39. Julia Steinberger, "Cogs in the Climate Machine," Climate Conscious (blog), July 1, 2021, https://medium.com/climate-conscious/cogs-in-the-climate-machine-167cf16750dd.

40. carla bergman, ed., *Trust Kids!* (AK Press, 2022).

41. Hannah Arendt, *Eichmann in Jerusalem: A Report on the Banality of Evil* (Penguin, 2006).

42. Peter Sutoris, *Educating for the Anthropocene: Schooling and Activism in the Face of Slow Violence* (MIT Press, 2022).

43. Harvard Radcliffe Institute, "Climate Justice Universities: Another Education Is Possible, by Jennie C. Stephens," 2023, YouTube video, https://www.youtube.com/watch?v=Jz1VAXIkJBM.

44. Subcomandante Marcos, *The Zapatistas' Dignified Rage: Final Public Speeches of Subcommander Marcos,* ed. Nick Henck, trans. Henry Gales (AK Press, 2018).

CHAPTER 4

1. Paul Brodeur, "Annals of Chemistry: In the Face of Doubt," *The New Yorker,* June 1, 1986, https://www.newyorker.com/magazine/1986/06/09 /in-the-face-of-doubt.

2. Morgan McFall-Johnsen, "2 Climate Activists Got Kicked Out of the World's Biggest Earth-Science Conference for Protesting, and One Says the Association Is 'Silencing Scientists,'" Business Insider, 2022, https://www .businessinsider.com/climate-researchers-ejected-from-agu-fall-meeting-banner -activism-interview-2022-12.

3. Rose Abramoff, "I'm a Scientist Who Spoke Up about Climate Change. My Employer Fired Me," *The New York Times,* January 10, 2023, sec. Opinion, https://www.nytimes.com/2023/01/10/opinion/scientist-fired-climate- change-activism.html.

4. Fabian Dablander et al., "Climate Change Engagement of Scientists," *Nature Climate Change* 14 (2024): 1033–39, https://doi.org/10.1038/s41558- 024-02091-2.

5. Thierry et al., "'No Research on a Dead Planet': Preserving the Socio-Ecological Conditions for Academia," *Frontiers in Education* 8 (2023), https:// www.frontiersin.org/articles/10.3389/feduc.2023.1237076; Nathaniel Geiger and Janet K. Swim, "Climate of Silence: Pluralistic Ignorance as a Barrier to Climate Change Discussion," *Journal of Environmental Psychology* 47 (September 1, 2016): 79–90, https://doi.org/10.1016/j.jenvp.2016.05.002; Esther Michelsen Kjeldahl and Vincent F. Hendricks, "The Sense of Social Influence: Pluralistic Ignorance in Climate Change," *EMBO Reports* 19, no. 11 (November 2018): e47185, https://doi.org/10.15252/embr.201847185.

6. Upbeat, "Disavowal: Sally Weintrobe and the Psychoanalysis of Climate Change Denial," Institute of Psychoanalysis—British Psychological Society, 2016, https://psychoanalysis.org.uk/blog/disavowal-sally-weintrobe- and-the-psychoanalysis-of-climate-change-denial.

7. Thierry et al., "'No Research on a Dead Planet.'"

8. Robert K Merton, *The Sociology of Science: Theoretical and Empirical Investigations* (University of Chicago Press, 1973).

9. Samuel Finnerty, Jared Piazza, and Mark Levine, "Scientists' Identities Shape Engagement with Environmental Activism," *Communications Earth & Environment* 5, no. 1 (May 8, 2024): 1–13, https://doi.org/10.1038 /s43247-024-01412-9.

10. Donna Haraway, "Situated Knowledges: The Science Question in Feminism and the Privilege of Partial Perspective," *Feminist Studies* 14, no. 3 (1988): 575–99, https://doi.org/10.2307/3178066.

11. Linda Tuhiwai Smith, *Decolonizing Methodologies: Research and Indigenous Peoples* (Bloomsbury, 2021).

12. Boaventura de Sousa Santos, *The End of the Cognitive Empire: The Coming of Age of Epistemologies of the South* (Duke University Press, 2018); Boaventura de Sousa Santos, *Decolonising the University: The Challenge of Deep Cognitive Justice* (Cambridge Scholars, 2018).

13. Smith, *Decolonizing Methodologies: Research and Indigenous Peoples*.

14. Shawn Wilson, *Research Is Ceremony: Indigenous Research Methods* (Fernwood, 2020).

15. Robin Wall Kimmerer, *Braiding Sweetgrass: Indigenous Wisdom, Scientific Knowledge and the Teachings of Plants* (Milkweed Editions, 2013).

16. See, for example, Thomas Wells, "Academics Should Not Be Activists," *3 Quarks Daily* (blog), December 17, 2018, https://3quarksdaily.com/3quarksdaily/2018/12/academics-should-not-be-activists.html.

17. Edward W. Said, *Orientalism* (Penguin Books, 2019).

18. Scientist Rebellion, "Charlie Gardner: The University in Times of Climate and Ecological Crisis," 2022, YouTube video, https://www.youtube.com/watch?v=g4CZy7BDiWs; Fernando Racimo et al., "The Biospheric Emergency Calls for Scientists to Change Tactics," *eLife* 11 (November 7, 2022): e83292, https://doi.org/10.7554/eLife.83292.

19. Eber Hampton, "5—Memory Comes Before Knowledge: Research May Improve If Researchers Remember Their Motives," *Canadian Journal of Native Education* 21 (1995), https://doi.org/10.14288/cjne.v21i.195782.

20. Michael E. Soulé, "What Is Conservation Biology?," *BioScience* 35, no. 11 (1985): 727–34, https://doi.org/10.2307/1310054; Curt Meine, Michael Soulé, and Reed F. Noss, "'A Mission-Driven Discipline': The Growth of Conservation Biology," *Conservation Biology* 20, no. 3 (2006): 631–51, https://doi.org/10.1111/j.1523-1739.2006.00449.x; Ezra Susser and Michaeline Bresnahan, "Origins of Epidemiology," *Annals of the New York Academy of Sciences* 954, no. 1 (2001): 6–18, https://doi.org/10.1111/j.1749-6632.2001.tb02743.x; Robert Cox, "Nature's 'Crisis Disciplines': Does Environmental Communication Have an Ethical Duty?," *Environmental Communication* 1, no. 1 (May 1, 2007): 5–20, https://doi.org/10.1080/17524030701333948.

21. Paul Feyerabend, *Science in a Free Society* (Verso Books, 2017).

22. Paul Weindling et al., "The Victims of Unethical Human Experiments and Coerced Research under National Socialism," *Endeavour* 40, no. 1 (March 2016): 1–6, https://doi.org/10.1016/j.endeavour.2015.10.005.

23. Nicholas D. Kristof, "Unmasking Horror—A Special Report; Japan Confronting Gruesome War Atrocity," *The New York Times,* March 17, 1995, sec. World, https://www.nytimes.com/1995/03/17/world/unmasking-horror-a -special-report-japan-confronting-gruesome-war-atrocity.html.

24. BBC News, "US Scientists 'Knew Guatemala Syphilis Tests Unethical,'" BBC News, August 29, 2011, sec. Latin America & Caribbean, https:// www.bbc.com/news/world-latin-america-14712089.

25. Fiona Harvey, "Universities Must Reject Fossil Fuel Cash for Climate Research, Say Academics," *The Guardian,* March 21, 2022, sec. Science, https://www.theguardian.com/science/2022/mar/21/universities-must-reject-fossil-fuel-cash-for-climate-research-say-academics.

26. Wells, "Academics Should Not Be Activists."

27. Tobias Brosch, "Affect and Emotions as Drivers of Climate Change Perception and Action: A Review," *Current Opinion in Behavioral Sciences,* Human Response to Climate Change: From Neurons to Collective Action, 42 (December 1, 2021): 15–21, https://doi.org/10.1016/j.cobeha.2021.02.001.

28. Notable for their absence are any depictions of bottom-up grassroots organizing, beyond a social media campaign by the scientists themselves telling people to "just look up." The rise in popular discontent at the impending doom is manifested in the form of panicking masses in disarray, looting cities without any clear strategy or goal. The scientists end up talking to these masses as if they were children in need of awakening. Political ecologists Patrick Bresnihan and Naomi Millner rightly ask, "Where are the social movements in the film?" (Patrick Bresnihan and Naomi Millner, *All We Want Is the Earth: Land, Labour and Movements Beyond Environmentalism,* First Edition [Bristol University Press, 2023].)

29. Donna Lu, "'It Parodies Our Inaction': Don't Look Up, an Allegory of the Climate Crisis, Lauded by Activists," *The Guardian,* December 30, 2021, sec. Environment, https://www.theguardian.com/environment/2021/dec/30 /it-parodies-our-inaction-dont-look-up-an-allegory-of-the-climate-crisis-lauded-by-activists.

30. See, for example, Robert S. Young, "Opinion | A Scientists' March on Washington Is a Bad Idea," *The New York Times,* January 31, 2017, sec. Opinion, https://www.nytimes.com/2017/01/31/opinion/a-scientists-march-on-washington-is-a-bad-idea.html.

31. Deborah Runkle and Mark S. Frankel, "Advocacy in Science. Summary of a Workshop Convened by the American Association for the Advancement of Science" (Workshop on Advocacy in Science, 2012); Augustin Fragnière, "The Public Engagement of Academics: From Academic Freedom to Professional Ethics—Report of the Working Group on Research and

Engagement" (University of Lausanne, 2022), https://www.unil.ch/files/live/sites/centre-durabilite/files/pdf/report-public-engagement-of-academics.pdf.

32. John E. Kotcher et al., "Does Engagement in Advocacy Hurt the Credibility of Scientists? Results from a Randomized National Survey Experiment," *Environmental Communication* 11, no. 3 (May 4, 2017): 415–29, https://doi.org/10.1080/17524032.2016.1275736.

33. Lindsey Beall et al., "Controversy Matters: Impacts of Topic and Solution Controversy on the Perceived Credibility of a Scientist Who Advocates," *PLOS One* 12, no. 11 (November 14, 2017): e0187511, https://doi.org/10.1371/journal.pone.0187511.

34. Viktoria Cologna et al., "Majority of German Citizens, US Citizens and Climate Scientists Support Policy Advocacy by Climate Researchers and Expect Greater Political Engagement," *Environmental Research Letters* 16, no. 2 (January 2021): 024011, https://doi.org/10.1088/1748-9326/abd4ac.

35. Pamela Pietrucci and Leah Ceccarelli, "Scientist Citizens: Rhetoric and Responsibility in L'Aquila," *Rhetoric and Public Affairs* 22, no. 1 (March 1, 2019): 95–128, https://doi.org/10.14321/rhetpublaffa.22.1.0095; Alice Benessia and Bruna De Marchi, "When the Earth Shakes . . . and Science with It. The Management and Communication of Uncertainty in the L'Aquila Earthquake," *Futures*, Post-Normal Science in Practice, 91 (August 1, 2017): 35–45, https://doi.org/10.1016/j.futures.2016.11.011.

36. Keynyn Brysse et al., "Climate Change Prediction: Erring on the Side of Least Drama?," *Global Environmental Change* 23, no. 1 (February 1, 2013): 327–37, https://doi.org/10.1016/j.gloenvcha.2012.10.008.

37. Jeff Tollefson, "Top Climate Scientists Are Sceptical That Nations Will Rein in Global Warming," *Nature* 599, no. 7883 (November 1, 2021): 22–24, https://doi.org/10.1038/d41586-021-02990-w.

38. Arielle Samuelson, "Why Did This Climate Scientist Chain Herself to a Pipeline?," Heated, July 12, 2021, https://heated.world/p/why-did-this-climate-scientist-chain.

39. Erin Stone, "Climate Scientists Chained Themselves to a Downtown Bank's Doors in an Act of Peaceful Protest. Police in Riot Gear Shut It Down," LAist, April 7, 2022, https://laist.com/news/climate-environment/climate-revolution-now-how-an-atypical-scientific-protest-led-to-arrests-in-dtla.

40. Kye Askins, "'That's Just What I Do': Placing Emotion in Academic Activism," *Emotion, Space and Society,* Activism and Emotional Sustainability, 2, no. 1 (July 1, 2009): 4–13, https://doi.org/10.1016/j.emospa.2009.03.005.

CHAPTER 5

1. Isabelle Stengers, *Another Science Is Possible: A Manifesto for Slow Science* (John Wiley & Sons, 2018).

2. Laura Paddison, "Germany Plans to Destroy This Village for a Coal Mine. Thousands Are Gathering to Stop It," CNN, 2023, https://edition.cnn.com/2023/01/14/europe/lutzerath-germany-coal-protests-climate-intl/index.html.

3. Lothar Becker, "So ist die Stimmung in Lützerath kurz vor der Räumung," January 8, 2023, https://www.zdf.de/uri/de44bbb3-eafe-4e6b-8ac1-03b78a3ffd15.

4. Katherine Tangalakis-Lippert, "Greta Thunberg and a 'Mud Wizard' Faced off against German Cops to Protest a Coal Deal with the Country's Largest Energy Company," Business Insider, 2023, https://www.businessinsider.com/climate-activists-protest-expanding-coal-pit-mud-wizard-lutzerath-2023-1.

5. Peter Gelderloos, *The Solutions Are Already Here: Strategies for Ecological Revolution from Below* (Pluto Press, 2022), https://doi.org/10.2307/j.ctv28vb1wq.

6. Stefan Collini, *What Are Universities For?* (Penguin UK, 2012).

7. Fernando Racimo et al., "Ethical Publishing: How Do We Get There?," *Philosophy, Theory, and Practice in Biology* 14 (October 21, 2022), https://doi.org/10.3998/ptpbio.3363.

8. Mohsen Kayal, Jane Ballard, and Ehsan Kayal, "Transformative Choices towards a Sustainable Academic Publishing System," *Ideas in Ecology and Evolution* 14 (January 27, 2022), https://doi.org/10.24908/iee.2021.14.3.f.

9. David Loher et al., "On Politics and Precarity in Academia," *Social Anthropology* 27, no. S2 (2019): 97–117, https://doi.org/10.1111/1469-8676.12695.

10. Daniel S. Katz et al., "Cost Models for Running an Online Open Journal," *Journal of Open Source Software Blog* (blog), April 6, 2019, https://blog.joss.theoj.org/2019/06/cost-models-for-running-an-online-open-journal; Samuel Alizon, "Inexpensive Research in the Golden Open-Access Era," *Trends in Ecology & Evolution* 33, no. 5 (May 1, 2018): 301–303, https://doi.org/10.1016/j.tree.2018.02.005; Alexander Grossmann and Björn Brembs, "Current Market Rates for Scholarly Publishing Services," *F1000Research* 10 (July 1, 2021): 20, https://doi.org/10.12688/f1000research.27468.2.

11. Vincent Larivière, Stefanie Haustein, and Philippe Mongeon, "The Oligopoly of Academic Publishers in the Digital Era," *PLOS One* 10, no. 6 (June 10, 2015): e0127502, https://doi.org/10.1371/journal.pone.0127502; The Guardian, "Editorial: The Guardian View on Academic Publishing: Disastrous Capitalism," *The Guardian*, March 4, 2019, sec. Opinion, https://

www.theguardian.com/commentisfree/2019/mar/04/the-guardian-view-on
-academic-publishing-disastrous-capitalism.

12. Robert Lazenby, "Time to Break Academic Publishing's Stranglehold on Research" *New Scientist*, November 21, 2018, https://www.newscientist.com /article/mg24032052-900-time-to-break-academic-publishings-stranglehold -on-research/.

13. Shaun Yon-Seng Khoo, "Article Processing Charge Hyperinflation and Price Insensitivity : An Open Access Sequel to the Serials Crisis," *LIBER Quarterly*, 29 (LIBER, 2019), 1–18, https://doi.org/10.18352/lq.10280.

14. Juliet Nabyonga-Orem et al., "Article Processing Charges Are Stalling the Progress of African Researchers: A Call for Urgent Reforms," *BMJ Global Health* 5, no. 9 (September 2020): e003650, https://doi.org/10.1136 /bmjgh-2020-003650.

15. Grossmann and Brembs, "Current Market Rates for Scholarly Publishing Services"; Larivière, Haustein, and Mongeon, "The Oligopoly of Academic Publishers in the Digital Era."

16. Racimo et al., "Ethical Publishing: How Do We Get There?"

17. Robert Hassan, "When Innovation Becomes Conformist: Academic Research in Network Time," in *Universities in the Flux of Time* (Routledge, 2014).

18. Sheila Slaughter and Larry L. Leslie, *Academic Capitalism: Politics, Policies, and the Entrepreneurial University* (Johns Hopkins University Press, 1999); Sheila Slaughter and Gary Rhoades, *Academic Capitalism and the New Economy: Markets, State, and Higher Education* (Johns Hopkins University Press, 2004).

19. Daniel Lee Kleinman and Steven P. Vallas, "Science, Capitalism, and the Rise of the 'Knowledge Worker': The Changing Structure of Knowledge Production in the United States," *Theory and Society* 30, no. 4 (2001): 451–92.

20. Racimo et al., "Ethical Publishing: How Do We Get There?"

21. Robert Morse, "Weighing the Efficiency of Highly Ranked Universities," *US News and World Report*, December 19, 2013. https://www.yahoo .com/news/weighing-efficiency-highly-ranked-universities-154403346.html.

22. Kleinman and Vallas, "Science, Capitalism, and the Rise of the 'Knowledge Worker.'"

23. Decca Aitkenhead, "Peter Higgs: I Wouldn't Be Productive Enough for Today's Academic System," *The Guardian*, December 6, 2013, sec. Science, https://www.theguardian.com/science/2013/dec/06/peter-higgs-boson -academic-system.

24. Stengers, *Another Science Is Possible*.

25. Charlie J. Gardner et al., "From Publications to Public Actions: The Role of Universities in Facilitating Academic Advocacy and Activism in the Climate and Ecological Emergency," *Frontiers in Sustainability* 2 (2021), https://www.frontiersin.org/article/10.3389/frsus.2021.679019.

26. Jennifer Richter et al., "Tempered Radicalism and Intersectionality: Scholar-Activism in the Neoliberal University," *Journal of Social Issues* 76, no. 4 (2020): 1014–35, https://doi.org/10.1111/josi.12401.

27. Sandra J. Grey, "Activist Academics: What Future?," *Policy Futures in Education* 11, no. 6 (December 1, 2013): 700–711, https://doi.org/10.2304 /pfie.2013.11.6.701.

28. Université de Fribourg, "Academia & Activism in a Time of Climate and Ecological Crises: Learning from the Past, Planning for the Future," Agenda—Universié de Fribourg, 2022, https://www.unifr.ch/scimed/en /agenda?eventid=12105.

29. Julian Steiner, "Free-speech—Das Wort hat Sanja Hakala," *Alma&Georges* (blog), April 6, 2023, https://www.unifr.ch/alma-georges /articles/2023/free-speech-das-wort-hat-sanja-hakala.

30. Farida Khali, "Free-speech—La parole à Arnaud Chiolero," *Alma&Georges* (blog), March 30, 2023, https://www.unifr.ch/alma-georges /articles/2023/free-speech-la-parole-a-arnaud-chiolero; Julian Steiner, "Free-speech—Das Wort hat Ivo Wallimann-Helmer" *Alma&Georges* (blog), April 13, 2023, https://www.unifr.ch/alma-georges/articles/2023/free-speech-das -wort-hat-ivo-wallimann-helmer; Christian Doninelli, "Free-speech—La parole à Aurianne Stroude," *Alma&Georges* (blog), July 25, 2023, https:// www.unifr.ch/alma-georges/articles/2023/free-speech-la-parole-a-aurianne-stroude; Farida Khali, "Free-speech—La parole à Robin Jolissaint," *Alma&Georges* (blog), March 2, 2023, https://www.unifr.ch/alma-georges /articles/2023/free-speech-la-parole-a-robin-jolissaint.

31. Scientist Rebellion, "SR Talks: How I Quit Neuroscience to Focus on Preventing Climate Breakdown, by Prof. Adam Aron," 2022, YouTube video, https://www.youtube.com/watch?v=EYYjH6reHwo.

32. 52 minutes RTS, "L'entretien de La Rédaction: Anaïs Tilquin, Porte-Parole d'Extinction Rebellion," 2021, YouTube video, https://www.youtube .com/watch?app=desktop&v=6CaRk7fgVKk; 52 minutes RTS, "Face à Face: Anaïs Tilquin," 2021, YouTube video, https://www.youtube.com/watch?v= KRj98oyjTak; Metabolism of Cities, "La Désobéissance Civile Pour Sauver La Planète (Podcast Avec Anaïs Tilquin—Renovate Switzerland)," 2022, YouTube video, https://www.youtube.com/watch?v=iEGdo3vW_W4.

33. Farzana Bashiri, "Conceptualizing Scholar-Activism Through Scholar-Activist Accounts," in *Making Universities Matter: Collaboration, Engagement, Impact,* ed. Pauline Mattsson, Eugenia Perez Vico, and Linus Salö

(Springer Nature Switzerland, 2024), 61–97, https://doi.org/10.1007/978-3-031-48799-6_4.

34. Paul Routledge and Kate Driscoll Derickson, "Situated Solidarities and the Practice of Scholar-Activism," *Environment and Planning D: Society and Space* 33, no. 3 (June 1, 2015): 391–407, https://doi.org/10.1177/0263775815594308.

35. Kye Askins, "'*That's Just What I Do*': Placing Emotion in Academic Activism," *Emotion, Space and Society,* Activism and Emotional Sustainability, 2, no. 1 (July 1, 2009): 4–13, https://doi.org/10.1016/j.emospa.2009.03.005; Stephen John Quaye, Mahauganee Dawn Shaw, and Dominique C. Hill, "Blending Scholar and Activist Identities: Establishing the Need for Scholar Activism," *Journal of Diversity in Higher Education* 10, no. 4 (2017): 381–99, https://doi.org/10.1037/dhe0000060.

36. Ajay Singh Chaudhary, *The Exhausted of the Earth: Politics in a Burning World* (Repeater, 2024).

CHAPTER 6

1. Stefano Harney and Fred Moten, *The Undercommons: Fugitive Planning and Black Study* (Autonomedia / Minor Compositions, 2013), https://www.minorcompositions.info/wp-content/uploads/2013/04/undercommons-web.pdf.

2. Nikoline Borgermann et al., "Debat: DTU Offshore støtter fossilindustrien. Vi bør fokusere på udfasning og ikke udvinding," Information, December 1, 2022, https://www.information.dk/debat/2022/12/dtu-offshore-stoetter-fossilindustrien-boer-fokusere-paa-udfasning-udvinding.

3. DTU, "Sustainability and Technology for Sustainable Development," August 23, 2023, https://baeredygtighed.dtu.dk/en/.

4. TotalEnergies, "TotalEnergies and the Technical University of Denmark Create a Center of Excellence in Decarbonized Energies," TotalEnergies .com, 2022, https://totalenergies.com/media/news/press-releases/research-totalenergies-and-technical-university-denmark-create-center.

5. Dr Paul Griffin, "The Carbon Majors Database: CDP Carbon Majors Report 2017," Climate Accountability Institute, 2017, https://cdn.cdp.net/cdp-production/cms/reports/documents/000/002/327/original/Carbon-Majors-Report-2017.pdf?1501833772.

6. TotalEnergies, "Natural Gas, a Key Energy Source for the Energy Transition," TotalEnergies.com, April 16, 2024, https://totalenergies.com/company/transforming/multi-energy-offer/natural-gaz.

7. Sahara Reporters, "TotalEnergies Oil Pollution Causing Deaths in Nigeria as Host Communities Forced to Engage in Illegal Crude Oil

Mining," 2023, Sahara Reporters, https://saharareporters.com/2023/03/25 /totalenergies-oil-pollution-causing-deaths-nigerian-state-host-communities -forced-engage.

8. Fred Pearce, "A Major Oil Pipeline Project Strikes Deep at the Heart of Africa," Yale E360, 2020, https://e360.yale.edu/features/a-major-oil-pipeline-project-strikes-deep-at-the-heart-of-africa.

9. Damian Carrington, "'Monstrous' East African Oil Project Will Emit Vast Amounts of Carbon, Data Shows," *The Guardian,* October 27, 2022, sec. Environment, https://www.theguardian.com/environment/2022/oct/27 /east-african-crude-oil-pipeline-carbon.

10. Divest DTU, "DTU Offshore Contract," Divest DTU, 2022, https:// divestdtu.dk/posts/dhrtc-contract/.

11. Scientist Rebellion Denmark, "Video of DTU Action," 2022, https:// twitter.com/SciRebDenmark/status/1593314699126493186?s=20.

12. Morten Larsen, "DTU-Forskere og studerende protesterer mod oliepenge," Arbejderen, November 18, 2022, https://arbejderen.dk/indland /forskere-og-studerende-protesterer-over-dtu-kontrakt-med-oliefirma/.

13. Rasmus Larsen, "Debat: DTU Offshores forskning understøtter for-syningssikkerheden under den grønne omstilling," Information, December 8, 2022, https://www.information.dk/debat/2022/12/dtu-offshores-forskning -understoetter-forsyningssikkerheden-groenne-omstilling.

14. Divest DTU, "DTU must stop taking fossil fuel money!," Divest DTU, 2022, https://divestdtu.dk/.

15. Nick Engelfried, "The Fossil Free Research Movement Is Taking Universities by Storm," Waging Nonviolence, October 3, 2022, https:// wagingnonviolence.org/2022/10/the-fossil-free-research-movement-is -taking-universities-by-storm/.

16. Douglas Almond, Xinming Du, and Anna Papp, "Favourability towards Natural Gas Relates to Funding Source of University Energy Centres," *Nature Climate Change* 12, no. 12 (2022): 1122–28, https://doi.org /10.1038/s41558-022-01521-3.

17. Phoebe Keane, "How the Oil Industry Made Us Doubt Climate Change," BBC, September 19, 2020, https://www.bbc.com/news/stories-53640382.

18. Sofia Hiltner et al., "Fossil Fuel Industry Influence in Higher Education: A Review and a Research Agenda," *WIREs Climate Change* 15, no. 6 (2024): e904, https://doi.org/10.1002/wcc.904.

19. Sebastião Vieira de Freitas Netto et al., "Concepts and Forms of Greenwashing: A Systematic Review," *Environmental Sciences Europe* 32, no. 1 (2020): 19, https://doi.org/10.1186/s12302-020-0300-3.

20. DTU Offshore, "About Us," August 23, 2023, https://offshore.dtu.dk /about-us.

21. DTU Offshore, "New Research Areas Cause Name Change for Research Centre—DTU," https://www.dtu.dk, 2022, https://www.dtu.dk/english/news /all-news/nyhed?id=65456e59-bc05-4c38-b8ee-0694db5a1007.

22. Divest DTU, "DTU Offshore Contract."

23. BBC, "Cambridge's Trinity College Lawn Dug up by Extinction Rebellion," BBC, 2020, https://www.bbc.com/news/uk-england-cambridgeshire-51534446.

24. Gemma Gardner, "Extinction Rebellion Cambridge Action at Trinity College, Cambridge," 2020, YouTube video, https://www.youtube.com /watch?v=Pe9SxTsiyWM.

25. Max Colbert, "UK Universities Take £40m in Fossil Fuel Funding since 2022," *The Guardian,* October 4, 2023, sec. Environment, https:// www.theguardian.com/environment/2023/oct/04/uk-universities-take-41m-in-fossil-fuel-funding-since-2022.

26. Geoffrey Supran and Benjamin Franta, "The Fossil Fuel Industry's Invisible Colonization of Academia," *The Guardian,* March 13, 2017, sec. Environment, https://www.theguardian.com/environment/climate-consensus-97-per-cent/2017/mar/13/the-fossil-fuel-industrys-invisible-colonization-of-academia; Jenna Corderoy, "British Universities Slammed for Taking £90m from Oil Companies in Four Years," openDemocracy, 2021, https://www .opendemocracy.net/en/dark-money-investigations/british-universities -slammed-for-taking-90m-from-oil-companies-in-four-years/.

27. Colbert, "UK Universities Take £40m in Fossil Fuel Funding since 2022."

28. TotalEnergies, "TotalEnergies and the Technical University of Denmark Create a Center of Excellence in Decarbonized Energies."

29. Holly Jean Buck, *Ending Fossil Fuels: Why Net Zero Is Not Enough* (Verso, 2021).

30. Laura Paddison, "The 'World's Largest' Vacuum to Suck Climate Pollution out of the Air Just Opened. Here's How It Works," CNN, May 8, 2024, https://www.cnn.com/2024/05/08/climate/direct-air-capture-plant -iceland-climate-intl/index.html.

31. Sara Budinis and Luca Lo Re, "Unlocking the Potential of Direct Air Capture: Is Scaling Up through Carbon Markets Possible?—Analysis," IEA, May 11, 2023, https://www.iea.org/commentaries/unlocking-the-potential-of-direct-air-capture-is-scaling-up-through-carbon-markets-possible.

32. Mark Poynting, "World's First Year-Long Breach of Key 1.5C Warming Limit," BBC, February 8, 2024, https://www.bbc.com/news/science-environment-68110310.

33. Alexandra Köves, "Can a Sustainability Transition Do Justice to the Global South—Roland Ngam," Economics for Rebels, December 31, 2023, https://open.spotify.com/show/47QC99TJF855AXozj4ShfX.

34. Joan Martínez-Alier, *Land, Water, Air and Freedom: The Making of World Movements for Environmental Justice* (Edward Elgar, 2023), https://www.e-elgar.com/shop/gbp/land-water-air-and-freedom-9781035312764.html.

35. ActionAid International, "'Our Lives Will Never Be the Same', Say Niger Delta Communities Devastated by Shell Gas Flares," ActionAid International, November 23, 2023, https://actionaid.org/news/2023/our-lives -will-never-be-same-say-niger-delta-communities-devastated-shell-gas-flares.

36. End Fossil, "End Fossil—Occupy!," 2023, https://endfossil.com/.

37. Emma Howard, "Shell in Deal with Top University to Influence Its Curriculum," Truthout, May 24, 2017, https://truthout.org/articles/revealed -shell-in-deal-with-top-university-to-influence-its-curriculum/.

38. Vatan Hüzeir and Germain Fraser, "A Pipeline of Ideas," Changerism, 2017, https://changerism.com/wp-content/uploads/2015/09/A-Pipeline-of-Ideas.pdf.

39. EcoWatch, "How This Oil Giant Influences Curriculum at Top Dutch University," EcoWatch, May 16, 2017, https://www.ecowatch.com/shell -erasmus-university-2409844294.html.

40. Hoger Onderwijs Persbureau, "Occupation Protest at UvA over Ties to Shell," Erasmus Magazine, January 17, 2023, https://www.erasmusmagazine .nl/en/2023/01/17/occupation-protest-at-uva-over-ties-to-shell/.

41. Erasmus Magazine, "Protest at Sanders Building," Erasmus Magazine, February 7, 2023, https://www.erasmusmagazine.nl/en/specials/protest-at -sanders-building/.

42. Elmer Smaling, "Erasmus Trust Fund Does Invest in Fossil Industries," Erasmus Magazine, December 20, 2022, https://www.erasmusmagazine.nl /en/2022/12/20/erasmus-trust-fund-does-invest-in-fossil-industries/.

43. Willem Schinkel, "In Praise of Flatness. On Campus Protest and Academic Community," Erasmus Magazine, February 22, 2023, https://www .erasmusmagazine.nl/en/2023/02/22/in-praise-of-flatness-on-campus-protest-and-academic-community/.

44. Fiona Harvey, "Universities Must Reject Fossil Fuel Cash for Climate Research, Say Academics," *The Guardian,* March 21, 2022, sec. Science, https://www.theguardian.com/science/2022/mar/21/universities-must-reject -fossil-fuel-cash-for-climate-research-say-academics.

45. Engelfried, "The Fossil Free Research Movement Is Taking Universities by Storm."

46. Engelfried.

47. Ilana Cohen, "A Dutch University Just Set a Powerful Precedent for Climate Research," *The Nation,* May 10, 2023, https://www.thenation.com /article/environment/fossil-free-research-policy-vrije-universiteit-amsterdam-climate-change/.

48. Stephen Burgen, "Barcelona Students to Take Mandatory Climate Crisis Module from 2024," *The Guardian,* November 12, 2022, sec. World News, https://www.theguardian.com/world/2022/nov/12/barcelona-students-to-take-mandatory-climate-crisis-module-from-2024.

49. Alaina Kinol et al., "Climate Justice in Higher Education: A Proposed Paradigm Shift towards a Transformative Role for Colleges and Universities," *Climatic Change* 176, no. 2 (2023): 15, https://doi.org/10.1007/s10584-023-03486-4.

50. Amy Westervelt, "Revealed: Leading Climate Research Publisher Helps Fuel Oil and Gas Drilling," *The Guardian,* 2022, https://www.theguardian.com/environment/2022/feb/24/elsevier-publishing-climate-science-fossil-fuels.

51. Elsevier, "Corporate Responsibility & Sustainability" www.elsevier.com, 2024, https://www.elsevier.com/about/sustainability.

52. Elsevier, "Elsevier's Climate Action Report 2021–2022," 2022, https://assets.ctfassets.net/o78em1y1w4i4/1osLLKHBRfxNIQBPcHGNBO/c22ff944368a9bcd6540b1e1654abc8d/Elsevier-Climate-Action-Report.pdf.

53. Stop Elsevier, "About Elsevier," *StopElsevier* (blog), February 22, 2023, https://stopelsevier.wordpress.com/more-info/.

54. Elsevier, "Knovel," 2024, https://app.knovel.com/kn.

55. Elsevier, "Geofacets User Guide," 2014, https://www.chest.ac.uk/media/1254/geofacets-user-guide-2014-web-version.pdf.

56. Open Secrets, "RELX Group Profile: Summary," OpenSecrets, 2024, https://www.opensecrets.org/orgs/relx-group/summary?id=D000067394.

57. Stop Elsevier, "About Elsevier."

58. Elsevier, "Elsevier's Climate Action Report 2021–2022."

59. Scientists for XR, "Why Would a Science and Healthcare Publisher Also Promote Climate Collapse? Today Scientists Infiltrate the @RELXHQ AGM to Ask Hard Questions about @ElsevierConnect's Business @TheLancet @richardhorton1 Https://T.Co/b3I4KRymBT," Tweet, Twitter, April 20, 2023, https://twitter.com/ScientistsX/status/1649076756035383298.

60. Tom Stafford, "Lessons from the Campaign against Elsevier 'We Won, but How Did We Win?,'" *ACME: An International Journal for Critical Geographies* 8, no. 3 (2009): 494–504.

61. Racimo et al., "Ethical Publishing: How Do We Get There?," *Philosophy, Theory, and Practice in Biology* 14 (October 21, 2022), https://doi.org/10.3998/ptpbio.3363.

62. Climate Rights Coalition, "Addressing Elsevier's Climate-Related Human Rights Impacts," Climate Rights Coalition, 2024, https://www.climaterightscoalition.com/.

63. Chantal Mouffe, *Agonistics: Thinking the World Politically* (Verso, 2013).

64. Schinkel, "In Praise of Flatness. On Campus Protest and Academic Community."

65. Schinkel.

CHAPTER 7

1. Lao-Tse, *The Tao Teh King* (Project Gutenberg, 2018), https://www.gutenberg.org/files/216/216-h/216-h.htm.

2. Thomas More, *Utopia* (Robert Roberts, 1878), https://ia801007.us.archive.org/11/items/utopia00more_4/utopia00more_4.pdf.

3. Symphony Private Jets, "Fly like You Own It," 2023, https://x.com/symphonyjets/status/1683907540554219550.

4. Gianluca Grimalda, "Refusing to Fly Has Lost Me My Job as a Climate Researcher. It's a Price Worth Paying," *The Guardian*, October 12, 2023, sec. Opinion, https://www.theguardian.com/commentisfree/2023/oct/12/fly-climate-breakdown-germany-climate-change-papua-new-guinea.

5. Gianluca Grimalda, "My #SlowTravel from Germany to PNG," Tweet, *Twitter*, February 16, 2023, https://twitter.com/GGrimalda/status/1626344390196068352.

6. The Syrian Observatory for Human Rights, "Syrian Revolution 13 Years on: Nearly 618,000 Persons Killed since the Onset of the Revolution in March 2011," *SOHR* (blog), March 15, 2024, https://www.syriahr.com/en/328044/.

7. UNHCR, "Syria Emergency," UNHCR, 2023, https://www.unhcr.org/emergencies/syria-emergency.

8. Rana Rahimpour, "Fury in Iran as Young Woman Dies Following Morality Police Arrest," BBC, September 16, 2022, https://www.bbc.com/news/world-middle-east-62930425.

9. Gianluca Grimalda, "Looking at Slums through the Train Window Is a Punch in the Stomach," Tweet, *Twitter*, March 18, 2023, https://twitter.com/GGrimalda/status/1637058063579357187.

10. Gianluca Grimalda, "I Visited 11 Villages Already in the Island of Bougainville, PNG, Researching Community Adaptation to Climate Change and Globalisation," Tweet, *Twitter*, June 18, 2023, https://twitter.com/GGrimalda/status/1670440361226031106.

11. Thorfinn Stainforth, "Linking Aviation Emissions to Climate Justice," *IEEP AISBL* (blog), 2019, https://ieep.eu/news/linking-aviation-emissions-to-climate-justice/.

12. Stay Grounded, "Aviation: A Matter of Climate Justice," Stay Grounded, 2023, https://stay-grounded.org/wp-content/uploads/2021/11/SG-Climate-Justice-and-Aviation-Factsheet.pdf.

13. Isabell Burian, "It Is Up in the Air: Academic Flying of Swedish Sustainability Academics and a Pathway to Organisational Change" (Master Thesis Series in Environmental Studies and Sustainability Science, Lund University, 2018), https://lup.lub.lu.se/luur/download?func=downloadFile&recordOId=8947780&fileOId=8949826; Lisa Jacobson, "The Virus and the Elephant in the Room: Knowledge, Emotions and a Pandemic—Drivers to Reducing Flying in Academia," in *Academic Flying and the Means of Communication*, ed. Kristian Bjørkdahl and Adrian Santiago Franco Duharte (Springer Nature, 2022), 209–35, https://doi.org/10.1007/978-981-16-4911-0_9; Nicholas A. Poggioli and Andrew J. Hoffman, "Decarbonising Academia's Flyout Culture," in *Academic Flying and the Means of Communication*, ed. Kristian Bjørkdahl and Adrian Santiago Franco Duharte (Springer Nature, 2022), 237–67, https://doi.org/10.1007/978-981-16-4911-0_10.

14. Helen E. Fox et al., "Why Do We Fly? Ecologists' Sins of Emission," *Frontiers in Ecology and the Environment* 7, no. 6 (2009): 294–96.

15. A. Stohl, "The Travel-Related Carbon Dioxide Emissions of Atmospheric Researchers," *Atmospheric Chemistry and Physics* 8, no. 21 (2008): 6499–6504, https://doi.org/10.5194/acp-8-6499-2008.

16. Our World in Data, "Per Capita CO_2 Emissions from International Passenger Flights, Tourism-Adjusted," Our World in Data, 2018, https://ourworldindata.org/grapher/per-capita-co2-international-flights-adjusted?tab=chart&country=ESP~USA~GBR~ARE~CAN~AUS~NOR~OWID_WRL.

17. Kristian Bjørkdahl and Adrian Santiago Franco Duharte, eds., *Academic Flying and the Means of Communication* (Springer Nature, 2022), https://doi.org/10.1007/978-981-16-4911-0.

18. Andreas Malm, *Fossil Capital: The Rise of Steam Power and the Roots of Global Warming* (Verso, 2016).

19. All European Academies, *Towards Climate Sustainability of the Academic System in Europe and Beyond*, (ALLEA, 2022), https://allea.org/portfolio-item/towards-climate-sustainability-of-the-academic-system-in-europe-and-beyond/; Andrew Glover, Yolande Strengers, and Tania Lewis, "The Unsustainability of Academic Aeromobility in Australian Universities," *Sustainability: Science, Practice and Policy* 13, no. 1 (2017): 1–12, https://doi.org/10.1080/15487733.2017.1388620.

20. Lorraine Whitmarsh et al., "Use of Aviation by Climate Change Researchers: Structural Influences, Personal Attitudes, and Information

Provision," *Global Environmental Change* 65 (2020): 102184, https://doi
.org/10.1016/j.gloenvcha.2020.102184.

21. Seth Wynes and Simon D. Donner, "Addressing Greenhouse Gas Emissions from Business-Related Air Travel at Public Institutions: A Case Study of the University of British Columbia" (Pacific Institute for Climate Solutions, 2018), https://pics.uvic.ca/sites/default/files/AirTravelWP_FINAL.pdf.

22. Joachim Ciers et al., "Carbon Footprint of Academic Air Travel: A Case Study in Switzerland," *Sustainability* 11, no. 1 (2019): 80, https://doi
.org/10.3390/su11010080.

23. All European Academies, *Towards Climate Sustainability of the Academic System in Europe and Beyond*.

24. Iselin Medhaug, "The ETH Zurich Flight Emission Project: Emissions 2016–2019" (ETH Zurich, 2021), https://doi.org/10.3929/ethz-b-000498721.

25. Julien Arsenault et al., "The Environmental Footprint of Academic and Student Mobility in a Large Research-Oriented University," *Environmental Research Letters* 14, no. 9 (2019): 095001, https://doi.org/10.1088
/1748-9326/ab33e6.

26. Agnes S. Kreil, "Does Flying Less Harm Academic Work? Arguments and Assumptions about Reducing Air Travel in Academia," *Travel Behaviour and Society* 25 (2021): 52–61, https://doi.org/10.1016/j.tbs.2021.04.011; Melissa Nursey-Bray et al., "The Fear of Not Flying: Achieving Sustainable Academic Plane Travel in Higher Education Based on Insights from South Australia," *Sustainability* 11, no. 9 (2019): 2694, https://doi.org/10.3390
/su11092694; Catherine Grant, "Academic Flying, Climate Change, and Ethnomusicology: Personal Reflections on a Professional Problem," *Ethnomusicology Forum* 27, no. 2 (2018): 123–35, https://doi.org/10.1080/17411912.
2018.1503063.

27. Martine Schaer, Cédric Jacot, and Janine Dahinden, "Transnational Mobility Networks and Academic Social Capital among Early-Career Academics: Beyond Common-Sense Assumptions," *Global Networks* 21, no. 3 (2021): 585–607, https://doi.org/10.1111/glob.12304.

28. Konstantinos Chalvatzis and Peter L. Ormosi, "The Carbon Impact of Flying to Economics Conferences: Is Flying More Associated with More Citations?," *Journal of Sustainable Tourism*, August 19, 2020, https://www
.tandfonline.com/doi/abs/10.1080/09669582.2020.1806858; Seth Wynes et al., "Academic Air Travel Has a Limited Influence on Professional Success," *Journal of Cleaner Production* 226 (2019): 959–67, https://doi.org/10.1016
/j.jclepro.2019.04.109.

29. Olivier Berné et al., "The Carbon Footprint of Scientific Visibility," *Environmental Research Letters* 17, no. 12 (2022): 124008, https://doi
.org/10.1088/1748-9326/ac9b51.

30. Stay Grounded and PCS, "A Rapid and Just Transition of Aviation— Shifting towards Climate-Just Mobility" (Stay Grounded, 2021), https:// stay-grounded.org/just-transition/.

31. Scientist Rebellion, *Scientists and Activists Block Private Airports in Eleven Countries*, 2022, YouTube video, https://www.youtube.com/watch?v= 2A0-y3WRQF8.

32. Andrew Murphy and Valentin Simon, "Private Jets: Can the Super Rich Supercharge Zero-Emission Aviation?," Transport & Environment, 2021, https://www.transportenvironment.org/wp-content/uploads/2021/05 /202209_private_jets_FINAL_with_addendum.pdf.

33. Chuck Collins, Omar Ocampo, and Kalena Thomhave, "High Flyers 2023: How Ultra-Rich Private Jet Travel Costs the Rest of Us and Burns Up Our Planet" (Institute for Policy Studies & Patriotic Millionaires, 2023).

34. Mira Alestig et al., "Carbon Inequality Kills: Why Curbing the Excessive Emissions of an Elite Few Can Create a Sustainable Planet for All," (Oxfam, 2024).

35. Sarah Jacob, Olivia Rudgard, and April Roach, "This Dutch Airport Wants to Ban Private Jets," *TIME*, May 3, 2023, https://time.com/6276617 /schiphol-amsterdam-airport-ban-private-jets/.

36. Cathy Buyck, "Eindhoven Airport in the Netherlands Announces Private Jet Ban," *Aviation International News*, 2023, https://www.ainonline.com /aviation-news/business-aviation/2023-11-14/eindhoven-airport-announces -private-jet-ban-2026.

37. Paolo Casalis, *The Researcher* (Produzioni Fuorifuoco, 2024), Vimeo video, https://vimeo.com/ondemand/researcher.

38. Amitav Ghosh, *The Nutmeg's Curse: Parables for a Planet in Crisis* (University of Chicago Press, 2021); Robin Kimmerer, *Braiding Sweetgrass: Indigenous Wisdom, Scientific Knowledge and the Teachings of Plants* (Milkweed Editions, 2013).

39. Bruce C. Glavovic, Timothy F. Smith, and Iain White, "The Tragedy of Climate Change Science," *Climate and Development* 14, no. 9 (2022): 829– 33, https://doi.org/10.1080/17565529.2021.2008855.

40. Wolfgang Cramer et al., "Klimaforscher Entlassen Weil Er Nicht Fliegen Wollte: Dr. Grimalda Sollte Seine Forschung Fortsetzen Können!," innn. it, 2023, https://innn.it/klimaforscher-entlassen.

41. Paul Weindling et al., "The Victims of Unethical Human Experiments and Coerced Research under National Socialism," *Endeavour* 40, no. 1 (2016): 1–6, https://doi.org/10.1016/j.endeavour.2015.10.005.

42. Edward J. Larson, *Sex, Race, and Science: Eugenics in the Deep South* (Johns Hopkins University Press, 1996).

43. Liam Doherty, "Investment, Influence and Integrity: How Arms Industries Compromise Universities' Research Agendas," Rethinking Security, March 13, 2023, https://rethinkingsecurity.org.uk/2023/03/13/investment-influence-and-integrity-how-arms-industries-compromise-universities-research-agendas/.

44. Steven Epstein, *Impure Science: AIDS, Activism, and the Politics of Knowledge*, First Edition (University of California Press, 1996).

45. Kelly Moore, *Disrupting Science: Social Movements, American Scientists, and the Politics of the Military, 1945–1975*, (Princeton University Press, 2009), https://doi.org/10.1515/9781400823802.

46. Bertrand Russell et al., "The Russell-Einstein Manifesto," Pugwash, 1955, https://pugwash.org/1955/07/09/statement-manifesto/.

47. Charlie J. Gardner and James M. Bullock, "In the Climate Emergency, Conservation Must Become Survival Ecology," *Frontiers in Conservation Science* 2 (2021), https://www.frontiersin.org/articles/10.3389/fcosc.2021.659912; Fabian Dablander, "Understanding and Preventing Climate Breakdown: A Guide for Social and Behavioral Scientists" (PsyArXiv, March 31, 2023), https://doi.org/10.31234/osf.io/4uzjs; Michael W. Apple, "Challenging the Epistemological Fog: The Roles of the Scholar/Activist in Education," *European Educational Research Journal*, May 11, 2016, https://doi.org/10.1177/1474904116647732.

CHAPTER 8

1. Stuart Jeffries, "David Graeber Interview: 'So Many People Spend Their Working Lives Doing Jobs They Think Are Unnecessary,'" *The Guardian*, March 21, 2015, sec. Books, https://www.theguardian.com/books/2015/mar/21/books-interview-david-graeber-the-utopia-of-rules.

2. The extent to which true horizontalism is practiced in the climate movement—particularly in Extinction Rebellion (XR)—has been recently challenged by Berglund and Schmidt: "We argue that XR is much less horizontalist than many other social movements. What XR calls a self-organising system is not based on the power of the assembly as is the case in horizontalist organisations. It is instead a more network-based model that is intended to disperse power in a transparent way and make for more efficient decision-making." See Oscar Berglund and Daniel Schmidt, *Extinction Rebellion and Climate Change Activism: Breaking the Law to Change the World* (Springer Nature, 2020).

3. Extinction Rebellion, "Affinity Group Welcome Pack," 2019, https://rebellion.earth/wp-content/uploads/2019/04/XR-Affinity-Group-Pack.pdf; Marianne Maeckelbergh, *The Will of the Many : How the Alterglobalisation*

Movement Is Changing the Face of Democracy (Pluto Press, 2009), https://cir.nii.ac.jp/crid/1130282271494980480.

4. Aric McBay, *Full Spectrum Resistance, Volume One: Building Movements and Fighting to Win* (Seven Stories Press, 2019).

5. Kerstine Appunn and Benjamin Wehrmann, "What Are the Climate and Energy Plans of German Election Winner Scholz?," Clean Energy Wire, September 29, 2021, https://www.cleanenergywire.org/news/what-are-climate-and-energy-plans-german-election-winner-scholz.

6. European Environment Agency, "Fossil Fuel Subsidies," November 17, 2023, https://www.eea.europa.eu/en/analysis/indicators/fossil-fuel-subsidies.

7. CBS News, "Germany's Olaf Scholz Urges World Health Summit to Continue after Protesters Pull Fire Alarm," 2022, YouTube video, https://www.youtube.com/watch?v=q3hjQj_kmM4.

8. Scientist Rebellion, *Scientist Rebellion Rings the Alarm at the World Health Summit,* 2022, YouTube video, https://www.youtube.com/watch?v=JNVJvx2iVBU.

9. Laurence Cox, "Scholarship and Activism: A Social Movements Perspective," *Studies in Social Justice* 9, no. 1 (December 10, 2015): 34–53, https://doi.org/10.26522/ssj.v9i1.1153.

10. María Isabel Casas-Cortés, Michal Osterweil, and Dana E. Powell, "Blurring Boundaries: Recognizing Knowledge-Practices in the Study of Social Movements," *Anthropological Quarterly* 81, no. 1 (2008): 17–58.

11. Cox, "Scholarship and Activism."

12. Wolfsburg.de, "Stadt Wolfsburg—Stadtchronik," Wolfsburg.de, August 10, 2023, https://www.wolfsburg.de/leben/stadtportraitstadtgeschichte/stadtchronik.

13. Autostadt GmbH, "Autostadt Completes Successful Year in 2017 with a Record Number of Visitors in a Single Month," PRNewswire, 2018, https://www.prnewswire.com/news-releases/autostadt-completes-successful-year-in-2017-with-a-record-number-of-visitors-in-a-single-month-669907643.html.

14. Yahoo News, "Scientists Rebellion Activists Glue Their Hands to the Porsche Exhibition Centre's Floor," Yahoo News, October 21, 2022, https://uk.news.yahoo.com/scientists-rebellion-activists-glue-hands-163333699.html.

15. M. Lange, M. Hendzlik, and M. Schmied, "Klimaschutz Durch Tempolimit: Wirkung Eines Generellen Tempolimits Auf Bundesautobahnen Auf Die Treibhausgasemissionen," *Umweltbundesamt,* no. 38 (2020), https://trid.trb.org/view/1894004.

16. Deutsche Umwelthilfe, "Klima Schützen, Leben Retten, Fossile Abhängigkeit Reduzieren: Tempolimit 80 Außerorts Und 100 Auf Autobahnen," 2020,

https://www.duh.de/fileadmin/user_upload/download/Projektinformation
/Verkehr/2019_Tempolimit/Hintergrundpapier_Tempolimit_2022_final.pdf.

17. Greenpeace, "VW Vor Gericht," 2022, https://www.greenpeace.de
/publikationen/VW_vor_Gericht_0.pdf; Simon Hage and Gerald Traufetter,
"Die letzte Schlacht um die Autoabgase," *Der Spiegel,* June 3, 2021, sec.
Wirtschaft, https://www.spiegel.de/wirtschaft/unternehmen/schadstoffnorm-
euro-7-die-letzte-schlacht-um-die-autoabgase-a-806fa6b0-f5ad-4897-96ba-
7e90f9f8c579.

18. Larena Klöckner, "9-Euro-Ticket und Lindner: Unfair für Reiche,"
Die Tageszeitung: taz, August 8, 2022, sec. Gesellschaft, https://taz.de
/!5870327/.

19. Der Spiegel, "Bilanz des Verbands Deutscher Verkehrsunternehmen:
9-Euro-Ticket hat positiven Klimaeffekt," *Der Spiegel,* August 29, 2022,
sec. Mobilität, https://www.spiegel.de/auto/9-euro-ticket-spart-fast-zwei
-millionen-tonnen-co-ein-a-cabaabcf-e7e7-4914-850f-daa9e820c64b.

20. Gianluca Grimalda, "Video of Autostadt Security Service," 2022,
https://twitter.com/GGrimalda/status/1584649601055068160.

21. Autostadt, "Statement Zur Protestaktion in Der Autostadt," 2022,
https://presse.autostadt.de/#954846/statement-zur-protestaktion.

22. Scientist Rebellion, "Volkswagen's Greenwashing and Lobbying Is
Killing People | Dr. Marta Matos," 2022, YouTube video, https://www
.youtube.com/watch?v=49Q-XWT8gMI.

23. HotNews.ro, "Încălzirea globală: Obiectivul de 1,5°C este mort, avert-
izează oamenii de știință / „Dați 100 de miliarde de dolari/an țărilor sărace","
2022, https://www.hotnews.ro/stiri-mediu-25869029-incalzirea-globala-
obiectivul-1-5c-este-mort-avertizeaza-oamenii-stiinta-dati-100-miliarde
-doalari-tarilor-sarace.htm; L'Obs, "Allemagne : des scientifiques militant pour
le climat laissés 42 heures collés au sol, dans le noir et sans chauffage," L'Obs,
October 25, 2022, https://www.nouvelobs.com/monde/20221025.OBS65103
/allemagne-des-scientifiques-militant-pour-le-climat-laisses-42-heures-colles
-au-sol-dans-le-noir-et-sans-chauffage.html; Giorgia Colucci, "Clima, Nove
Attivisti 'incollati' Alle Auto Di Lusso Di Porsche in Germania: 'Volkswagen
Ritarda Le Sue Azioni Contro Il Riscaldamento Globale,'" Il Fatto Quotidiano,
2022, https://www.ilfattoquotidiano.it/2022/10/21/clima-nove-attivisti-
incollati-alle-auto-di-lusso-di-porsche-in-germania-volkswagen-ritarda-le-
sue-azioni-contro-il-riscaldamento-globale-video/6846924/; 324cat, "Scientist
Rebellion ocupa un centre de Porsche per exigir límit de velocitat a Alemanya,"
CCMA, October 21, 2022, https://www.ccma.cat/324/scientist-rebellion-
ocupa-un-centre-de-porsche-per-exigir-limit-de-velocitat-a-alemanya
/noticia/3191392/; Independent TV, "Scientists Rebellion Activists Glue
Themselves to Porsche Showroom," Independent TV, 2022, https://www

.independent.co.uk/tv/news/scientist-rebellion-climate-change-germany-b2208138.html; Die Welt, "Wolfsburg: Klimaaktivisten kleben sich in Autostadt fest—und beklagen, dass Heizung und Licht ausgehen," Die Welt, October 21, 2022, https://www.welt.de/vermischtes/article241720585/Wolfsburg-Klimaaktivisten-kleben-sich-in-Autostadt-fest-und-beklagen-dass-Heizung-und-Licht-ausgehen.html.

24. Giuliana Miranda, "Crise do clima torna protestos de ambientalistas mais radicais," Folha de S. Paulo, November 5, 2022, https://www1.folha.uol.com.br/ambiente/2022/11/crise-do-clima-torna-protestos-de-ambientalistas-mais-radicais.shtml; 김한솔, "'밤샘시위 추운데 난방 왜 껐냐'... 포르쉐 쇼룸 장악한 환경단체의 황당한 요구사항," 인사이트, October 25, 2022, https://www.insight.co.kr/news/416381; Leo Lagos, "Investigadores llaman a la desobediencia civil de científicos y científicas para ayudar a revertir la emergencia ecológica," la diaria, December 3, 2022, https://ladiaria.com.uy/ciencia/articulo/2022/12/investigadores-llaman-a-la-desobediencia-civil-de-cientificos-y-cientificas-para-ayudar-a-revertir-la-emergencia-ecologica/; Halk TV, "İklim aktivistlerinden Porsche protestosu: Ellerini yapıştırdılar," Halk TV, October 20, 2022, https://halktv.com.tr/cevre/iklim-aktivistlerinden-porsche-protestosu-ellerini-yapistirdilar-698787h.

25. Oscar Berglund, "Disruptive Protest, Civil Disobedience & Direct Action," Politics, 2023, https://journals.sagepub.com/doi/full/10.1177/02633957231176999.

26. Daniele Artico et al., "'Beyond Being Analysts of Doom': Scientists on the Frontlines of Climate Action," Frontiers in Sustainability 4 (2023), https://www.frontiersin.org/articles/10.3389/frsus.2023.1155897.

27. Extern, "Blutiger Protest: 25 Wissenschaftler demonstrieren gegen VW-Lobbyismus," regionalHeute.de, October 19, 2022, https://regionalheute.de/wolfsburg/blutiger-protest-25-wissenschaftler-demonstrieren-gegen-vw-lobbyismus-wolfsburg-1666188644/; Peter Carstens, "Klimaproteste bei VW und BMW: Warum Forschende ihre Komfortzone verlassen," geo.de, November 2, 2022, https://www.geo.de/natur/nachhaltigkeit/klimaproteste—warum-forscher-ihre-komfortzone-verlassen-32873358.html; Deutschlandfunk Kultur, "Scientist Rebellion—sollen Naturgesetze Demokratie aushebeln?," Deutschlandfunk Kultur, 2022, https://www.deutschlandfunkkultur.de/scientist-rebellion-sollen-naturgesetze-demokratie-aushebeln-dlf-kultur-3cf0f93c-100.html.

CHAPTER 9

1. Cistem Failure, "A Bigger Cage Is Still a Prison," 2017, https://cistemfailure.bandcamp.com/track/a-bigger-cage-is-still-a-prison.

2. Scientist Rebellion, "Scientist Rebellion Action at Munich BMW Welt," 2022, YouTube video, https://www.youtube.com/watch?v=69LKEU7A jDU&themeRefresh=1.

3. v.r., "Encarcelados 15 científicos en Alemania por su protesta por la inacción climática," La Opinión de A Coruña, November 1, 2022, https://www.laopinioncoruna.es/sociedad/2022/11/01/encarcelados-15-cientificos-alemania-protesta-77982037.html; Frank Jordan and Christine Kerler, "Polizei nimmt 16 Klimaaktivisten mehrere Tage in Gewahrsam," BR24, October 31, 2022, https://www.br.de/nachrichten/bayern/polizei-nimmt-16-klimaaktivisten-mehrere-tage-in-gewahrsam,TLkohBP; Alberto Reineri, "La protesta degli studiosi del clima contro il greenwashing di BMW," November 3, 2022, https://www.bikeitalia.it/2022/11/03/la-protesta-degli-studiosi-del-clima-contro-il-greenwashing-di-bmw/; Thomas Baïeto, "Crise climatique : des scientifiques français emprisonnés en Allemagne pour des actions de désobéissance civile," Franceinfo, November 3, 2022, https://www.francetvinfo.fr/monde/environnement/crise-climatique/crise-climatique-des-scientifiques-francais-emprisonnes-en-allemagne-pour-des-actions-de-desobeissance-civile_5455537.html; Olivier Monod, "Quatre scientifiques français en garde à vue en Allemagne pour leur activisme climatique," Libération, 2022, https://www.liberation.fr/international/europe/quatre-scientifiques-francais-en-garde-a-vue-en-allemagne-pour-leur-activisme-climatique-20221102_HRSVO7KBDJGZXIAT3ZPAXLLL7E/; Antonio Albiñana, "La rebelión de los científicos," El Tiempo, November 3, 2022, https://www.eltiempo.com/opinion/columnistas/antonio-albinana/la-rebelion-de-los-cientificos-columna-de-antonio-albinana-714960.

4. Global Witness, "Standing Firm: The Land and Environmental Defenders on the Frontlines of the Climate Crisis," 2023, https:///en/campaigns/environmental-activists/standing-firm/.

5. Jason Hickel, "Degrowth: A Theory of Radical Abundance," Real-World Economics Review, no. 87 (2019), http://www.paecon.net/PAEReview/issue87/Hickel87.pdf; Carolyn Merchant, The Death of Nature: Women, Ecology, and the Scientific Revolution, Reprint edition (HarperOne, 1990).

6. Amitav Ghosh, The Nutmeg's Curse: Parables for a Planet in Crisis (University of Chicago Press, 2021).

7. Andreas Malm, Fossil Capital: The Rise of Steam Power and the Roots of Global Warming (Verso, 2016).

8. Peter D. Norton, "Street Rivals: Jaywalking and the Invention of the Motor Age Street," Technology and Culture 48, no. 2 (2007): 331–59.

9. Ulrich Brand and Markus Wissen, The Imperial Mode of Living (Verso, 2021).

10. Giulio Mattioli et al., "The Political Economy of Car Dependence: A Systems of Provision Approach," *Energy Research & Social Science* 66 (August 1, 2020): 101486, https://doi.org/10.1016/j.erss.2020.101486.

11. Center for Biological Diversity, "Center for Biological Diversity Calls for Protection of Atlanta Forest, Independent Probe of Activist Killing," Center for Biological Diversity, accessed August 16, 2024, https://biological-diversity.org/w/news/press-releases/center-for-biological-diversity-calls-for-protection-of-atlanta-forest-independent-probe-of-activist-killing-2023-02-23/.

12. Charles Bethea, "The New Fight Over an Old Forest in Atlanta," *The New Yorker*, August 3, 2022, https://www.newyorker.com/news/letter-from-the-south/the-new-fight-over-an-old-forest-in-atlanta.

13. Natasha Lennard, "The Crackdown on Cop City Protesters Is So Brutal Because of the Movement's Success," The Intercept, January 27, 2023, https://theintercept.com/2023/01/27/cop-city-atlanta-forest/.

14. Tess Owen, "Police Shot 'Stop Cop City' Activist 14 Times with Their Hands Up, Independent Autopsy Shows," *VICE* (blog), March 13, 2023, https://www.vice.com/en/article/cop-city-activist-shot-hands-up-tortuguita-death/.

15. News2Share, "'Elders Say: Stop Cop City' Protest 'Cop City' at Brasfield & Gorrie Construction Sites in Atlanta," 2023, YouTube video, https://www.youtube.com/watch?v=8NIXmLj7m_4.

16. David Graeber, *Bullshit Jobs: A Theory* (Penguin Books, 2019); Erik Gomez-Baggethun, "Rethinking Work for a Just and Sustainable Future," *Ecological Economics* 200 (October 1, 2022): 107506, https://doi.org/10.1016/j.ecolecon.2022.107506.

17. The KU Degrowth Network, "The Degrowth Online Seminar Series: 'Degrowth in practice: Rethinking work and working time' with Professor Erik Gomez-Baggethun from the Norwegian University of Life Sciences," Københavns Universitets Videoportal, 2023, https://video.ku.dk/video/85853901/the-degrowth-online-seminar-series.

18. André Gorz, "Métamorphoses du travail: quête du sens, critique de la raison économique," 1988, https://philpapers.org/rec/GORMDT.

19. Ajay Singh Chaudhary, *The Exhausted of the Earth: Politics in a Burning World* (Repeater, 2024).

20. Stefan Collini, *What Are Universities For?* (Penguin UK, 2012).

21. Graeber, *Bullshit Jobs: A Theory*.

22. Maggie Berg and Barbara K. Seeber, *The Slow Professor: Challenging the Culture of Speed in the Academy* (University of Toronto Press, 2016).

23. Fabian Dablander et al., "Climate Change Engagement of Scientists," *Nature Climate Change*, August 5, 2024, 1–7, https://doi.org/10.1038/s41558-024-02091-2.

24. Max Weber, *Economy and Society: An Outline of Interpretive Sociology* (University of California Press, 1978).

25. Charlie J. Gardner et al., "From Publications to Public Actions: The Role of Universities in Facilitating Academic Advocacy and Activism in the Climate and Ecological Emergency," *Frontiers in Sustainability* 2 (2021), https://www.frontiersin.org/article/10.3389/frsus.2021.679019.

26. Malm, *Fossil Capital: The Rise of Steam Power and the Roots of Global Warming.*

27. David Graeber and David Wengrow, *The Dawn of Everything: A New History of Humanity* (Penguin UK, 2021).

28. Government Office for Science, "Migration and Global Environmental Change," October 20, 2011, https://www.gov.uk/government/collections /migration-and-global-environmental-change.

29. Todd Miller, Nick Buxton, and Mark Akkerman, "Global Climate Wall: How the World's Wealthiest Nations Prioritize Borders over Climate Action," Transnational Institute, June 4, 2024, https://www.tni.org/en/publication /global-climate-wall.

30. UN News, "Climate Change Recognized as 'Threat Multiplier', UN Security Council Debates Its Impact on Peace," January 25, 2019, https:// news.un.org/en/story/2019/01/1031322.

31. Rashid Khalidi, *The Hundred Years' War on Palestine: A History of Settler Colonialism and Resistance, 1917–2017,* Illustrated edition (Metropolitan Books, 2020).

32. Omar Shakir, "A Threshold Crossed," Human Rights Watch, April 27, 2021, https://www.hrw.org/report/2021/04/27/threshold-crossed/israeli-authorities-and-crimes-apartheid-and-persecution.

33. Israel-Palestine Timeline, "Deaths and Injuries in Israel-Palestine since 2000," Israel-Palestine Timeline, May 4, 2024, https://israelpalestinetimeline .org/.

34. Jonathan Yerushalmy, "Crisis in Gaza: Why Food, Water and Power Are Running Out," *The Guardian,* October 17, 2023, sec. World news, https://www.theguardian.com/world/2023/oct/17/crisis-gaza-why-food-water-power-running-out.

35. Al-Haq and EWASH, "Israel's Violations of Human Rights Regarding Water and Sanitation in the OPT—Report by Al-Haq and EWASH to CESCR—Non-Un Document" (United Nations, 2011), https://www.un.org /unispal/document/auto-insert-195880/.

36. Israeli scholar Maya Wind has documented that academia plays an important role in this form of oppression too. Through military degree programs, weapons research, campus outposts in Palestinian lands, and the ruthless repression of student voices of dissent, Israeli universities serve as

pillars of support for the maintenance and expansion of subjugation and apartheid; see Maya Wind, *Towers of Ivory and Steel: How Israeli Universities Deny Palestinian Freedom* (Verso, 2024); United Nations High Commissioner for Human Rights, "A/HRC/55/72: Israeli Settlements in the Occupied Palestinian Territory, Including East Jerusalem, and in the Occupied Syrian Golan—Report of the United Nations High Commissioner for Human Rights," OHCHR, 2024, https://www.ohchr.org/en/documents/reports /ahrc5572-israeli-settlements-occupied-palestinian-territory-including-east.

37. Sanjana Karanth, "Israeli Defense Minister: 'We Are Fighting Human Animals,'" HuffPost, September 10, 2023, https://www.huffpost.com/entry /israel-defense-minister-human-animals-gaza-palestine_n_6524220ae4b09 f4b8d412e0a.

38. Amnesty International, "Israel's Apartheid against Palestinians: Cruel System of Domination and Crime against Humanity," Amnesty International, 2022, https://www.amnesty.org/en/documents/mde15/5141/2022 /en/.

39. Al Jazeera, "Fears of a Ground Invasion of Gaza Grow as Israel Vows 'Mighty Vengeance,'" Al Jazeera, 2023, https://www.aljazeera.com/news/2023 /10/7/world-is-watching-fears-grow-of-a-massive-gaza-invasion-by-israel.

40. Gözde Bayar, "Israeli Bombardment Destroys More than 70% of Civilian Infrastructure in Gaza: UN Agency," Anadolu Agency, 2024, https:// www.aa.com.tr/en/middle-east/israeli-bombardment-destroys-more-than -70-of-civilian-infrastructure-in-gaza-un-agency/3138876.

41. AJLabs, "The Human Toll of Israel's War on Gaza – By the Numbers," Al Jazeera, January 15, 2025, https://www.aljazeera.com/news/2025/1/15 /the-human-toll-of-israels-war-on-gaza-by-the-numbers.

42. Rasha Khatib, Martin McKee, and Salim Yusuf, "Counting the Dead in Gaza: Difficult but Essential," *The Lancet* 404, no. 10449 (July 20, 2024): 237–38, https://doi.org/10.1016/S0140-6736(24)01169-3.

43. Nour Abuaisha, "4,000 Amputations, 2,000 Spinal, Brain Injuries in Gaza Amid Israeli Ongoing Onslaught," Anadolu Ajansı, July 12, 2024, https://www.aa.com.tr/en/middle-east/4-000-amputations-2-000-spinal-brain-injuries-in-gaza-amid-israeli-ongoing-onslaught/3417631.

44. United Nations Office for the Coordination of Humanitarian Affairs, "Hostilities in the Gaza Strip and Israel—Reported Impact | Day 236," United Nations Office for the Coordination of Humanitarian Affairs—occupied Palestinian territory, May 29, 2024, http://www.ochaopt.org /content/hostilities-gaza-strip-and-israel-reported-impact-day-236.

45. Francesca Albanese, "Anatomy of a Genocide—Report of the Special Rapporteur on the Situation of Human Rights in the Palestinian Territory Occupied since 1967 to Human Rights Council—Advance Unedited Version

(A/HRC/55/73)," United Nations, 2024, https://www.un.org/unispal/document/anatomy-of-a-genocide-report-of-the-special-rapporteur-on-the-situation-of-human-rights-in-the-palestinian-territory-occupied-since-1967-to-human-rights-council-advance-unedited-version-a-hrc-55/.

46. International Court of Justice, "Summary of the Order of 26 January 2024," 2024, https://www.icj-cij.org/node/203454.

47. Al-Haq, "Genocide Scholars and 100 Palestinian and International Civil Society Organisations Call on Prosecutor Khan to Issue Arrest Warrants, Investigate Israeli Crimes and Intervene to Deter Incitement to Commit Genocide in Gaza," Al-Haq | Defending Human rights in Palestine since 1979, 2024, https://www.alhaq.org/advocacy/21946.html.

48. International Rescue Committee, "The Collapse of Gaza's Health System," May 6, 2024, https://www.rescue.org/article/collapse-gazas-health-system.

49. Ghazal Golshiri, "All 12 Universities in Gaza Have Been the Target of Israeli Attacks: 'It's a War against Education,'" *Le Monde.Fr*, March 7, 2024, https://www.lemonde.fr/en/international/article/2024/03/07/all-12-universities-in-gaza-have-been-the-target-of-israeli-attacks-it-s-a-war-against-education_6592965_4.html; UN Office of the High Commissioner for Human Rights, "UN Experts Deeply Concerned over 'Scholasticide' in Gaza," OHCHR, 2024, https://www.ohchr.org/en/press-releases/2024/04/un-experts-deeply-concerned-over-scholasticide-gaza.

50. Médecins sans frontières, "Strikes, Raids and Incursions: Seven Months of Relentless Attacks on Healthcare in Palestine," 2024, https://www.msf.org/strikes-raids-and-incursions-seven-months-relentless-attacks-healthcare-palestine.

51. The New Arab, "Israel Targets Scientific and Literary Elite in Gaza," The New Arab, January 23, 2024, https://www.newarab.com/features/israel-targets-scientific-and-literary-elite-gaza.

52. Sharon Wrobel, "Amid Ongoing War, BP and Eni among Firms Awarded Gas Exploration Licences in Israel," *The Times of Israel*, October 30, 2023, https://archive.ph/NUOaN; Rachel Donald, "Everybody Wants Gaza's Gas," *Planet: Critical* (blog), October 31, 2023, https://www.planetcritical.com/p/everybody-wants-gazas-gas.

53. Ben Samuels and Amir Tibon, "U.S. to Push Israel on Allowing Gaza Offshore Gas Reserves to Revitalize Palestinian Economy—Israel News," *Haaretz*, November 20, 2023, https://www.haaretz.com/israel-news/2023-11-20/ty-article/.premium/u-s-to-push-israel-to-allow-gaza-offshore-gas-reserves-to-revitalize-palestinian-economy/0000018b-ed90-ddc3-afdb-fdd1ff250000.

54. UNCTAD, "The Economic Costs of the Israeli Occupation for the Palestinian People: The Unrealized Oil and Natural Gas Potential," August 22, 2019, https://unctad.org/system/files/official-document/gdsapp2019d1_en.pdf.

55. Benjamin Neimark et al., "A Multitemporal Snapshot of Greenhouse Gas Emissions from the Israel-Gaza Conflict," SSRN Scholarly Paper, January 5, 2024, https://doi.org/10.2139/ssrn.4684768.

56. Josie Glausiusz, "Israel Is Flooding Gaza's Tunnel Network: Scientists Assess the Risks," *Nature*, February 2, 2024, https://doi.org/10.1038/d41586-024-00320-4.

57. Eve Thomas, "Rebuilding Gaza: The Carbon Cost of War," *Energy Monitor* (blog), June 7, 2024, https://www.energymonitor.ai/news/rebuilding-gaza-the-carbon-cost-of-war/.

58. Hannah Brown, "Climate Protests in Madrid: 'The Red Paint Represents the Blood of All Those Who Have Died,'" Euronews, 2022, https://www.euronews.com/green/2022/04/07/climate-protests-in-madrid-the-red-paint-represents-the-blood-of-all-those-who-have-died.

59. J.L. Ferrer and Efe, "La Fiscalía pasa a considerar a 'Extinction Rebellion' y Futuro Vegetal como grupos 'terroristas,'" elperiodico, September 9, 2023, https://www.elperiodico.com/es/medio-ambiente/20230909/fiscalia-pasa-considerar-extinction-rebellion-91884518.

60. Andrés Actis, "Una policía se infiltra en organizaciones climáticas y el ecologismo pide explicaciones a Marlaska," 2023, https://www.lapoliticaonline.com/espana/politica-es/una-policia-se-infiltra-en-organizaciones-climaticas-y-el-ecologismo-pide-explicaciones-a-marlaska/.

61. El Salto Madrid, "Tres Activistas de Futuro Vegetal Detenidas Tras Pintar La Fachada de La Embajada Británica," El Salto, 2023, https://www.elsaltodiario.com/cambio-climatico/nueva-accion-futuro-vegetal-senala-fachada-embajada-britanica-madrid; Nacho Martín, "Por qué una novia de Futuro Vegetal tiñó de rojo la embajada británica," El Independiente, June 16, 2023, https://www.elindependiente.com/futuro/medio-ambiente/2023/06/17/la-novia-de-futuro-vegetal-que-tino-de-rojo-la-embajada-britanica-para-pedir-que-liberen-a-su-prometido/.

62. Scientist Rebellion, "Scientists on Trial," Scientist Rebellion, 2023, http://scientistrebellion.org/scientists-on-trial/.

CHAPTER 10

1. Farhana Sultana, "The Unbearable Heaviness of Climate Coloniality," *Political Geography* 99 (2022): 102638.

2. Jeff Masters, "A Record 63 Billion-Dollar Weather Disasters Hit Earth in 2023," Yale Climate Connections, January 18, 2024, http://yaleclimate connections.org/2024/01/a-record-63-billion-dollar-weather-disasters-hit-earth-in-2023/.

3. Andrew L. Fanning et al., "The Social Shortfall and Ecological Overshoot of Nations," *Nature Sustainability* 5, no. 1 (January 2022): 26–36, https://doi.org/10.1038/s41893-021-00799-z; Jason Hickel et al., "Imperialist Appropriation in the World Economy: Drain from the Global South through Unequal Exchange, 1990–2015," *Global Environmental Change* 73 (March 1, 2022): 102467, https://doi.org/10.1016/j.gloenvcha.2022.102467.

4. Sultana, "The Unbearable Heaviness of Climate Coloniality"; Tim Gore, "Confronting Carbon Inequality: Putting Climate Justice at the Heart of the COVID-19 Recovery," Oxfam, 2020; Yannick Oswald, Anne Owen, and Julia K Steinberger, "Large Inequality in International and Intranational Energy Footprints between Income Groups and across Consumption Categories," *Nature Energy* 5, no. 3 (2020): 231–39.

5. Naomi Klein, *This Changes Everything: Capitalism vs. The Climate* (Simon & Schuster, 2015).

6. XR Scotland, "Argentinian Activist Esteban on How to Build Climate Action in Global Solidarity," XR Scotland, 2021, https://xrscotland .org/2021/02/argentinian-activist-esteban-on-how-to-build-climate-action -in-global-solidarity/.

7. livialie, "Vaca Muerta Is a 'Carbon Bomb' That Could Eat Up More than 11% of the Global CO_2 Budget," 350, November 3, 2022, https://350 .org/vaca-muerta-is-a-carbon-bomb/; opsur, "Vaca Muerta Megaproject—A Fracking Carbon Bomb in Patagonia," Observatorio Petrolero Sur (blog), February 5, 2018, https://opsur.org.ar/2018/02/05/vaca-muerta-megaproject -a-fracking-carbon-bomb-in-patagonia/.

8. Henning Husum, "Total Stop for Greenwashing by Esteban Servat from Debt for Climate," 2023, YouTube video, https://www.youtube.com /watch?v=UiCdO-KyAv4.

9. Uki Goñi, "Indigenous Mapuche Pay High Price for Argentina's Fracking Dream," *The Guardian*, October 14, 2019, sec. Environment, https:// www.theguardian.com/environment/2019/oct/14/indigenous-mapuche -argentina-fracking-communities.

10. Joan Martínez-Alier, *Land, Water, Air and Freedom: The Making of World Movements for Environmental Justice* (Edward Elgar, 2023), https:// www.e-elgar.com/shop/gbp/land-water-air-and-freedom-9781035312764 .html; Maïa Courtois, "En Ouganda, Les Pressions à l'encontre Des Opposants de Total s'intensifient," Reporterre, 2021, https://reporterre.net/En-Ouganda -les-pressions-a-l-encontre-des-opposants-de-Total-s-intensifient.

11. Nina Lakhani, "'Very Disturbing': Crackdown on Oil Pipeline Protests in Uganda Concerns UN Rights Expert," *The Guardian*, October 19, 2023, sec. World News, https://www.theguardian.com/world/2023/oct/19/uganda-police-assault-arrest-oil-pipeline-protestors.

12. Nina Lakhani, "How Dash for African Oil and Gas Could Wipe Out Congo Basin Tropical Forests," *The Guardian*, November 10, 2022, sec. World News, https://www.theguardian.com/world/2022/nov/10/dash-african-gas-wipe-out-congo-basin-rainforests.

13. Josephine Moulds, "'We Won't Compromise': Villagers Rail against DRC's Fossil Fuel Auctions," The Bureau of Investigative Journalism, 2023, https://www.thebureauinvestigates.com/stories/2023-11-18/we-wont-compromise-villagers-rail-against-drcs-fossil-fuel-auctions.

14. Mike Mwenda, "DRC Oil Exploration in Protected Areas Draws Environmental Warnings," LifeGate Daily, 2023, https://www.lifegate.com/drc-oil-exploration-in-protected-areas-draws-environmental-warnings.

15. Adem Ay, "How Congolese Climate Activists Stopped a 'Carbon Bomb,' for Now," Waging Nonviolence, January 3, 2025, https://wagingnonviolence.org/2025/01/how-congolese-climate-activists-stopped-a-carbon-bomb-for-now/; Ministry of Hydrocarbons, Press release: Notice of cancellation of the tender process for the 27 oil blocks, October 11, 2024, https://hydrocarbures.gouv.cd/2024/10/14/communique-avis-annulation-processus-appel-doffres-des-27-blocs-petroliers/.

16. Don't Gas Africa, "Don't Gas Africa," Don't Gas Africa, 2023, https://dont-gas-africa.org.

17. Indigenous Environmental Network and Oil Change International, "Indigenous Resistance Against Carbon," 2021, https://www.ienearth.org/indigenous-resistance-against-carbon/.

18. Martínez-Alier, *Land, Water, Air and Freedom: The Making of World Movements for Environmental Justice.*

19. Arnim Scheidel et al., "Environmental Conflicts and Defenders: A Global Overview," *Global Environmental Change* 63 (July 1, 2020): 102104, https://doi.org/10.1016/j.gloenvcha.2020.102104.

20. Samantha Mailhot and Patricia E. Perkins, "Social Equity Is the Foundation of Degrowth," in *Degrowth & Strategy: How to Bring about Social-Ecological Transformation* (Mayfly, 2022).

21. Global Witness Report 2019, "Defending Tomorrow: The Climate Crisis and Threats against Land and Environmental Defenders," 2020, https://www.globalwitness.org/documents/19939/Defending_Tomorrow_EN_low_res_-_July_2020.pdf; Nathalie Butt et al., "The Supply Chain of Violence," *Nature Sustainability* 2 (2019): 742–47.

22. Margot S. Bass et al., "Global Conservation Significance of Ecuador's Yasuní National Park," *PLOS ONE* 5, no. 1 (January 19, 2010): e8767, https://doi.org/10.1371/journal.pone.0008767.

23. Martínez-Alier, *Land, Water, Air and Freedom: The Making of World Movements for Environmental Justice.*

24. Karen Osorio, "Yes to Yasuní: Democracy Survives Latest Threat in Ecuador," Rainforest Foundation US, August 28, 2023, https://rainforestfoundation.org/yes-to-yasuni/; Scientist Rebellion, "Scientist Rebellion—La Ciencia Es Clara," Mastodon, August 18, 2023, https://social.rebellion.global/@ScientistRebellion/110913126411231607.

25. Claudia Rebaza and Hannah Holland, "Ecuadorians Vote to Ban Oil Drilling in the Amazon in 'Historic' Referendum," CNN, 2023, https://edition.cnn.com/2023/08/21/americas/ecuador-oil-drilling-amazon-climate-intl/index.html; Steven Donziger, "The People of Ecuador Just Made Climate Justice History. The World Can Follow," *The Guardian*, August 31, 2023, sec. Opinion, https://www.theguardian.com/commentisfree/2023/aug/31/ecuador-oil-drilling-ban-climate-solution.

26. Helmut Haberl et al., "A Systematic Review of the Evidence on Decoupling of GDP, Resource Use and GHG Emissions, Part II: Synthesizing the Insights," *Environmental Research Letters* 15, no. 6 (June 2020): 065003, https://doi.org/10.1088/1748-9326/ab842a.

27. Martínez-Alier, *Land, Water, Air and Freedom: The Making of World Movements for Environmental Justice.*

28. Juan Zamorano, "Panama's Leader Calls for Referendum on Mining Concession, Seeking to Calm Protests over the Deal," AP News, October 31, 2023, https://apnews.com/article/panama-mining-environment-canada-aab0c46dd8a0dfb2930154b831a2e3ac.

29. Luke Taylor, "'Historic Moment': Panama Activists Celebrate Ruling against Copper Mine," *The Guardian*, November 28, 2023, sec. World News, https://www.theguardian.com/world/2023/nov/28/panama-supreme-court-canadian-copper-mine-unconstitutional.

30. Patrick Oppmann, "Two Demonstrators Killed amid Anti-Mining Protests in Panama," CNN, November 9, 2023, https://www.cnn.com/2023/11/08/americas/demonstrators-killed-mining-protests-panama/index.html; Mariela Laksman, "Panama Journalist Loses an Eye during an Environmental Protest When Police Shot into the Crowd," Orato, January 7, 2024, https://orato.world/2024/01/07/panama-journalist-loses-an-eye-during-an-environmental-protest-when-police-shot-into-the-crowd/.

31. Rosalba Icaza and Rolando Vázquez, "Diversity or Decolonisation? Researching Diversity at the University of Amsterdam," in *Decolonising the University* (Pluto Press, 2018).

32. Maegan Parker Brooks and Davis W. Houck (eds.), "'Nobody's Free Until Everybody's Free': Speech Delivered at the Founding of the National Women's Political Caucus, Washington, D.C., July 10, 1971," in *The Speeches of Fannie Lou Hamer: To Tell It Like It Is,* ed. Maegan Parker Brooks and Davis W. Houck (Jackson, MS, 2010) online edn., Mississippi Scholarship Online, March 20, 2014, https://academic.oup.com/mississippi-scholarship -online/book/29348/chapter-abstract/244099842.

CHAPTER 11

1. David Streitfeld, "Writing Nameless Things: An Interview with Ursula K. Le Guin," Los Angeles Review of Books, November 17, 2017, https:// lareviewofbooks.org/article/writing-nameless-things-an-interview-with -ursula-k-le-guin.

2. Helen Pidd, "Rules, Rotas and Revolutionary Song at Climate Action Camp," *The Guardian*, August 14, 2007, sec. Environment, https://www .theguardian.com/business/2007/aug/15/theairlineindustry.transportintheuk; Douglas Rogers, "Climate Camp Is Back—And It's Trying Something New," Novara Media, 2023, https://novaramedia.com/2023/07/04/climate-camp-is -back-and-its-trying-something-new/.

3. Calum McGeown and John Barry, "Agents of (Un)Sustainability: Democratising Universities for the Planetary Crisis," *Frontiers in Sustainability* 4 (2023), https://doi.org/10.3389/frsus.2023.1166642.

4. McGeown and Barry.

5. Blanca Missé and James Martel, "For Democratic Governance of Universities: The Case for Administrative Abolition," *Theory & Event* 27, no. 1 (2024): 5–29.

6. Missé and Martel.

7. Fabian Dablander et al., "Climate Change Engagement of Scientists," *Nature Climate Change,* 14 (2024): 1033–39, https://doi.org/10.1038 /s41558-024-02091-2.

8. Anne E. Urai and Clare Kelly, "Rethinking Academia in a Time of Climate Crisis," *eLife* 12 (2023): e84991, https://doi.org/10.7554/eLife.84991; Movement for a Free Academia, "The Gothenburg Manifesto for a Free Academia," 2024, https://www.freeacademia.org/.

9. Paulo Freire, *Pedagogy of the Oppressed* (Penguin UK, 2017).

10. Lisette van Beek, "Reimagining the University in the Climate Crisis: A Catalogue of Transformative University Practices in Northern Europe" (Utrecht University, 2024), https://www.uu.nl/sites/default/files/Reimagining- the-University-in-the-Climate-Crisis-LisetteVanBeek.pdf; Jennifer Atkinson and Sarah Jaquette Ray, eds., *The Existential Toolkit for Climate Justice*

Educators: How to Teach in a Burning World (University of California Press, 2024).

11. Max Koch, "Rethinking State-Civil Society Relations," in *Degrowth & Strategy: How to Bring about Social-Ecological Transformation* (Mayfly, 2022).

12. Sheila Slaughter and Larry L. Leslie, *Academic Capitalism: Politics, Policies, and the Entrepreneurial University* (Johns Hopkins University Press, 1999).

13. Sharon Stein, *Unsettling the University* (Johns Hopkins University Press, 2022), https://doi.org/10.56021/9781421445052.

14. Boaventura de Sousa Santos, *Decolonising the University: The Challenge of Deep Cognitive Justice* (Cambridge Scholars, 2018); Gurminder K. Bhambra, Dalia Gebrial, and Kerem Nişancıoğlu, *Decolonising the University* (Pluto Press, 2018); Ramón Grosfoguel, "The Structure of Knowledge in Westernized Universities: Epistemic Racism/Sexism and the Four Genocides/Epistemicides of the Long 16th Century," *Human Architecture: Journal of the Sociology of Self-Knowledge* 11, no. 1 (September 22, 2013), https://scholarworks.umb.edu/humanarchitecture/vol11/iss1/8.

15. Roy Macleod, "Passages in Imperial Science: From Empire to Commonwealth," *Journal of World History* 4, no. 1 (1993): 117–50; Frankie Chappell, "Terra Nullius?" The Royal Society (blog), April 20, 2020, https://royalsociety.org/blog/2020/04/terra-nullius/.

16. Peter Gelderloos, *The Solutions Are Already Here: Strategies for Ecological Revolution from Below* (Pluto Press, 2022).

17. Gelderloos.

18. Amitav Ghosh, *The Nutmeg's Curse: Parables for a Planet in Crisis* (University of Chicago Press, 2021).

19. Bhambra, Gebrial, and Nişancıoğlu, *Decolonising the University.*

20. Greta Thunberg, *No One Is Too Small to Make a Difference* (Penguin Random House, 2019).

21. Aaron Thierry et al., "'No Research on a Dead Planet': Preserving the Socio-Ecological Conditions for Academia," *Frontiers in Education* 8 (2023), https://www.frontiersin.org/articles/10.3389/feduc.2023.1237076.

22. Freire, *Pedagogy of the Oppressed.*

23. Fernando Racimo, "Academic Complicity in Times of Mass Murder," *University Post* (blog), 2024, https://uniavisen.dk/en/academic-complicity-in-times-of-mass-murder/; 29 Employees at the University of Copenhagen, "What the University of Copenhagen can learn from Rafah Garden," *University Post* (blog), June 7, 2024, https://uniavisen.dk/en/what-the-university-can-learn-from-rafah-garden/.

24. Jennie C. Stephens, *Climate Justice and the University: Shaping a Hopeful Future for All* (Johns Hopkins University Press, 2024).

25. McGeown and Barry, "Agents of (Un)Sustainability."

26. Linda Tuhiwai Smith, *Decolonizing Methodologies: Research and Indigenous Peoples* (Bloomsbury Publishing, 2021). Arturo Escobar, *Pluriversal Politics: The Real and the Possible* (Duke University Press, 2020).

27. I am indebted to H.L.T. Quan who—through her recent book on abolition feminism (H.L.T. Quan, *Become Ungovernable: An Abolition Feminist Ethic for Democratic Living* [London: Pluto Press, 2024])—led me to find this study.

28. Paola Bacchetta et al., "Queer of Color Space-Making in and beyond the Academic Industrial Complex," *Critical Ethnic Studies* 4, no. 1 (2018): 44–63, https://doi.org/10.5749/jcritethnstud.4.1.0044.

29. Boaventura de Sousa Santos, *Decolonising the University: The Challenge of Deep Cognitive Justice.*

30. Mike Emiliani, "In Rural Mexico, Student-Led Education Heals Old Wounds—YES! Magazine Solutions Journalism," *Yes Magazine,* 2013, https://www.yesmagazine.org/social-justice/2013/01/12/rural-mexico-student-led-education-heals-old-wounds-unitierra.

31. The Ecoversities Alliance, "Ecoversities—Reclaiming Knowledges, Relationships and Imaginations," Ecoversities, 2018, https://ecoversities.org/.

32. Stephen Jay Gould, *The Panda's Thumb: More Reflections in Natural History* (W.W. Norton & Company, 2010).

33. Peter Sutoris, *Educating for the Anthropocene: Schooling and Activism in the Face of Slow Violence* (MIT Press, 2022).

EPILOGUE

1. Carolyne L. Ryan, "The Conquest of the Desert and Argentina's Indigenous Peoples," in *Oxford Research Encyclopedia of Latin American History,* 2023, https://doi.org/10.1093/acrefore/9780199366439.013.1211; John Soluri, "The Conquest of the Desert: Argentina's Indigenous Peoples and the Battle for History," *Hispanic American Historical Review* 101, no. 4 (November 1, 2021): 719–21, https://doi.org/10.1215/00182168-9366844.

Selected Readings

INTRODUCTION

Berglund, Oscar, "Disruptive Protest, Civil Disobedience & Direct Action," *Politics* (2023), https://journals.sagepub.com/doi/full/10.1177/02633957231176999.

Gardner, Charlie J., and Claire F.R. Wordley, "Scientists Must Act on Our Own Warnings to Humanity," *Nature Ecology & Evolution* 3, no. 9 (2019): 1271–72.

CHAPTER 1

Brand, Ulrich, and Markus Wissen, *The Imperial Mode of Living* (Verso, 2021).

Capstick, Stuart, et al., "Civil Disobedience by Scientists Helps Press for Urgent Climate Action," *Nature Climate Change* 12 (2022): 773–74.

Hickel, Jason, *Less Is More: How Degrowth Will Save the World* (Windmill Books, 2021).

Malm, Andreas, *Fossil Capital: The Rise of Steam Power and the Roots of Global Warming* (Verso, 2016).

Schmelzer, Matthias, Andrea Vetter, and Aaron Vansintjan, *The Future Is Degrowth: A Guide to a World Beyond Capitalism* (Verso, 2022).

CHAPTER 2

Gardner, Charlie J., et al., "From Publications to Public Actions: The Role of Universities in Facilitating Academic Advocacy and Activism in the Climate and Ecological Emergency," *Frontiers in Sustainability* 2 (2021).

Nyberg, Daniel, Christopher Wright, and Vanessa Bowden, *Organising Responses to Climate Change: The Politics of Mitigation, Adaptation and Suffering* (Cambridge University Press, 2022).

Steinberger, Julia, "A Postmortem for Survival: On Science, Failure and Action on Climate Change," Age of Awareness (blog), May 8, 2019.

CHAPTER 3

bell hooks, *Teaching to Transgress: Education as the Practice of Freedom* (Routledge, 1994).

Freire, Paulo, *Pedagogy of the Oppressed* (Penguin Random House, 2017).

Racimo, Fernando, et al., "The Biospheric Emergency Calls for Scientists to Change Tactics," *eLife* 11 (2022): e83292.

Steinberger, Julia, "The Kids Are Not OK", Yale Climate Connections (blog), May 15, 2022.

CHAPTER 4

Dablander, Fabian, et al., "Climate Change Engagement of Scientists," *Nature Climate Change*, 14 (2024): 1033–39.

Haraway, Donna, "Situated Knowledges: The Science Question in Feminism and the Privilege of Partial Perspective," *Feminist Studies* 14, 3 (1988): 575–99.

Kimmerer, Robin, *Braiding Sweetgrass: Indigenous Wisdom, Scientific Knowledge and the Teachings of Plants* (Milkweed Editions, 2013).

Smith, Linda Tuhiwai, *Decolonizing Methodologies: Research and Indigenous Peoples* (Bloomsbury, 2021).

Wilson, Shawn, *Research Is Ceremony: Indigenous Research Methods* (Fernwood, 2020).

CHAPTER 5

Chaudhary, Ajay Singh, *The Exhausted of the Earth: Politics in a Burning World* (Repeater, 2024).

Collini, Stefan, *What Are Universities For?* (Penguin UK, 2012).

Larivière, Vincent, Stefanie Haustein, and Philippe Mongeon, "The Oligopoly of Academic Publishers in the Digital Era," *PLOS One* 10, no. 6 (2015): e0127502.

Racimo, Fernando, et al., "Ethical Publishing: How Do We Get There?," *Philosophy, Theory, and Practice in Biology* 14 (2022): 15.

Stengers, Isabelle, *Another Science Is Possible: A Manifesto for Slow Science* (John Wiley & Sons, 2018).

CHAPTER 6

Harney, Stefano, and Fred Moten, *The Undercommons: Fugitive Planning and Black Study* (Autonomedia / Minor Compositions, 2013).

Hiltner, Sofia, et al., "Fossil Fuel Industry Influence in Higher Education: A Review and a Research Agenda," *WIREs Climate Change* 15, no. 6 (2024): e904.

Schinkel, Willem, "In Praise of Flatness. On Campus Protest and Academic Community," Erasmus Magazine, February 22, 2023, https://www.erasmusmagazine.nl/en/2023/02/22/in-praise-of-flatness-on-campus-protest-and-academic-community/.

CHAPTER 7

Alestig, Mira, et al., "Carbon Inequality Kills: Why curbing the excessive emissions of an elite few can create a sustainable planet for all," Oxfam International, October 2024.

Gardner, Charlie J., and James M. Bullock, "In the Climate Emergency, Conservation Must Become Survival Ecology," *Frontiers in Conservation Science* 2 (2021).

Ghosh, Amitav, *The Nutmeg's Curse: Parables for a Planet in Crisis* (John Murray, 2022).

CHAPTER 8

Artico, Daniele, et al., "'Beyond Being Analysts of Doom': Scientists on the Frontlines of Climate Action," *Frontiers in Sustainability* 4 (2023).

Cox, Laurence, "Scholarship and Activism: A Social Movements Perspective," *Studies in Social Justice* 9, 1 (2015): 34–53.

CHAPTER 9

Graeber, David, *Bullshit Jobs: A Theory* (Penguin Books, 2019).

Khalidi, Rashid, *The Hundred Years' War on Palestine: A History of Settler Colonialism and Resistance, 1917–2017*, Illustrated edition (Metropolitan Books, 2020).

Merchant, Carolyn, *The Death of Nature: Women, Ecology, and the Scientific Revolution* (HarperOne, 1990).

Wind, Maya, *Towers of Ivory and Steel: How Israeli Universities Deny Palestinian Freedom* (Verso, 2024).

CHAPTER 10

Global Witness Report, "Defending Tomorrow: The Climate Crisis and Threats against Land and Environmental Defenders," Global Witness, 2020, https://www.globalwitness.org/en/campaigns/environmental-activists/defending-tomorrow.

Indigenous Environmental Network and Oil Change International, "Indigenous Resistance Against Carbon," 2021, https://www.ienearth.org/indigenous-resistance-against-carbon/.

Martínez-Alier, Joan, *Land, Water, Air and Freedom: The Making of World Movements for Environmental Justice* (Edward Elgar, 2023).

Sultana, Farhana, "The Unbearable Heaviness of Climate Coloniality," *Political Geography* 99 (2022): 102638.

CHAPTER 11

Bhambra, Gurminder K., Dalia Gebrial, and Kerem Nişancıoğlu (eds.), *Decolonising the University* (Pluto Press, 2018).

McGeown, Calum, and John Barry, "Agents of (Un)Sustainability: Democratising Universities for the Planetary Crisis," *Frontiers in Sustainability* 4 (2023).

Missé, Blanca, and James Martel, "For Democratic Governance of Universities: The Case for Administrative Abolition," *Theory & Event* 27, no. 1 (2024): 5–29.

Stephens, Jennie C., *Climate Justice and the University: Shaping a Hopeful Future for All* (Johns Hopkins University Press, 2024).

Thierry, Aaron, et al., "'No Research on a Dead Planet': Preserving the Socio-Ecological Conditions for Academia," *Frontiers in Education* 8 (2023).

Urai, Anne E., and Clare Kelly, "Rethinking Academia in a Time of Climate Crisis," *eLife* 12 (2023): e84991.

Index

Note: *f* indicates a figure or image.

Founded in 1893,
UNIVERSITY OF CALIFORNIA PRESS
publishes bold, progressive books and journals
on topics in the arts, humanities, social sciences,
and natural sciences—with a focus on social
justice issues—that inspire thought and action
among readers worldwide.

The UC PRESS FOUNDATION
raises funds to uphold the press's vital role
as an independent, nonprofit publisher, and
receives philanthropic support from a wide
range of individuals and institutions—and from
committed readers like you. To learn more, visit
ucpress.edu/supportus.